Breaking Inert Bonds:
Organic Synthesis

Shireen

– Table of Contents –

– Chapter One –

Introduction to Cp^{X}Rh(III)-catalyzed C–H Activation

1.1 Importance of C–H Functionalization

The direct conversion of carbon–hydrogen bonds into valuable carbon-carbon and carbon-heteroatom bonds is a significant challenge to synthetic organic chemists.[1] Carbon–Hydrogen bonds are the most common motifs in small molecules, and due to this ubiquity, this makes C–H bonds among the most desirable candidates to be manipulated and transformed into valuable targets. The high bond dissociation energy (BDE) presents the first challenge. Compared to that of pre-functionalized analogues such as aryl halides (Figure 1.1) C–H bonds are significantly more inert.

	R–H	R₂CH–H	R₃C–H	vinyl C–H	aryl C–H		aryl–I
BDE (kcal/mol)	98.2	95.1	93.2	110.0	110.9	>>	51
pKa	50	55	71	44	43		

Figure 1.1 Characteristics in the diversity of C–H bonds in organic synthesis.

When considering C–H bonds, their high pKa value compared to that of heteroatom–H bonds presents another challenge to consider. For these reasons, a method that has gained significant traction in the field of organic synthesis throughout the years is metal-catalyzed C–H bond activation.

1.2 Modes of C–H Activation

Predominately, there are two pathways to break a C–H bond and form a new metal-carbon bond as an intermediate.

Pathway 1: Oxidative Addition

Pathway 2: Concerted Metallation-Deprotonation (CMD)

Figure 1.2 *Modes of C–H bond activation to furnish new carbon-metal bonds.*

The first is oxidative addition, where the metal center is formally oxidized by 2 electrons to both a metal-carbon bond and a metal-hydride bond. The second pathway involves a ligand-assisted deprotonation event named concerted metallation-deprotonation (CMD). This pathway often involves weak bases with κ^2 binding modes (such as carboxylates, carbonates, phosphates, etc.) and is thought to occur in a redox-neutral concerted process.[2] The process relies on an agostic interaction to acidify the C–H bond, enabling a 6-membered transition state for deprotonation and subsequent metalation of the carbon unit. Each mode of reactivity provides advantages and disadvantages; but on the whole, CMD tends to be a much milder method to functionalize C–H bonds.

1.3 Installation of Directing Groups

While C–H bonds can be activated in a number of ways, the challenge of selectively cleaving a specific bond remains.

In order to combat this challenge, chemists install directing groups with heteroatoms that put the metal in close proximity to the C–H bond to be activated. This strategy relies heavily on confirmation of resulting metallacyclic species for further functionalization to occur.

Figure 1.3 *Selectivity challenges of C–H activation.*

Among other metals, rhodium(III) piano stool complexes bearing a cyclopentadienyl (Cp) ligand have shown great selectivity and diversity in functionalization methods in recent years. In association with a carboxylate-type base (shown in Figure 1.4 with an acetate ligand), the rhodium catalyst uses the

3 coordination sites beneath the Cp ligand to selectively convert a C–H bond into more important motifs such as C–C, C–N, C–O bonds.[3]

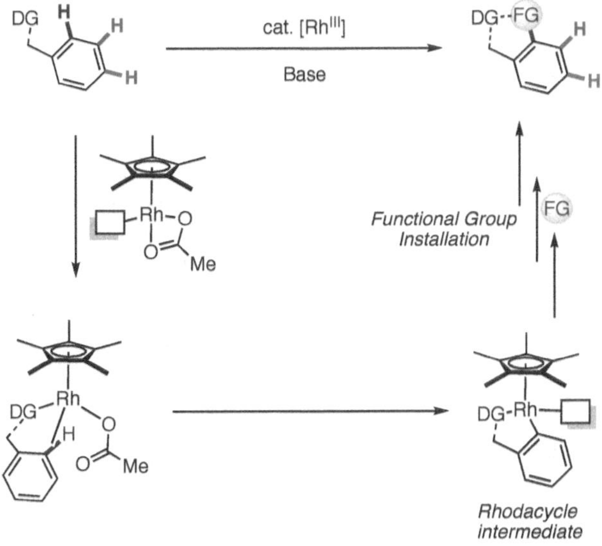

Figure 1.4 *Directing group assisted Rh(III)-catalyzed ortho-C–H functionalization.*

In 2010, our group joined the community and took advantage of this reactivity by treating secondary benzamides with alkynes in the presence of copper(II) acetate and a rhodium(III) catalyst bearing a pentamethylcyclopentadiene (Cp*) ligand.[4] First the dimer pre-catalyst is broken up to liberate the Rh-diacetato active catalyst. This species can deprotonate the N–H bond of the benzamide revealing a directing group toward the *ortho*-C–H bond. This complex undergoes C–H activation by a CMD type mechanism that gives rise to a 5-membered metallacycle. After migratory insertion affords the 7-membered metallacycle, reductive elimination forms a C–N bond and gives the isoquinolone product. Finally, 2 equivalents of Cu^{II} oxidize the resulting Rh^{I} species to regenerate the catalyst.

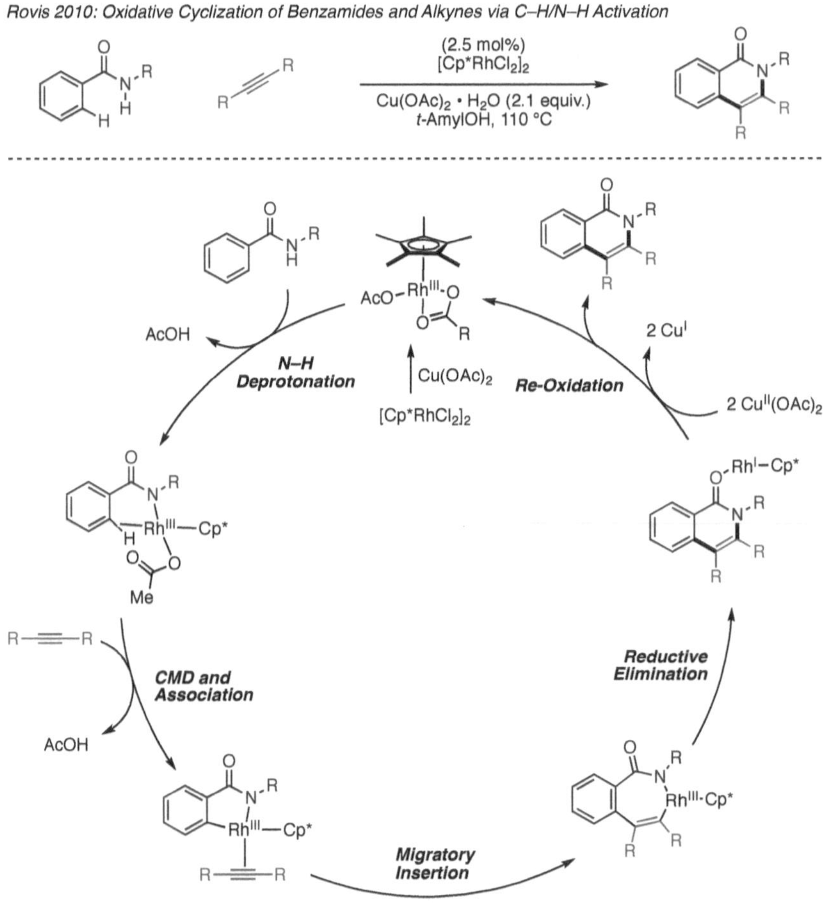

Figure 1.5 Mechanism of Rh(III)-catalyzed benzannulation of alkynes.

Around the same time, Fagnou and coworkers published a similar reaction with the installation of an internal oxidant as opposed to exogenous, stoichiometric amounts of copper(II) acetate.[5] This reactivity manifold allows for the same benzannulation to occur under milder conditions. Fagnou and coworkers then optimized the oxidative directing group from –OMe to –OPiv.[6] This alteration allows for the chemistry to happen at room temperature as well as expanding its scope to the insertion of alkenes, giving dihydroisoquinolones.

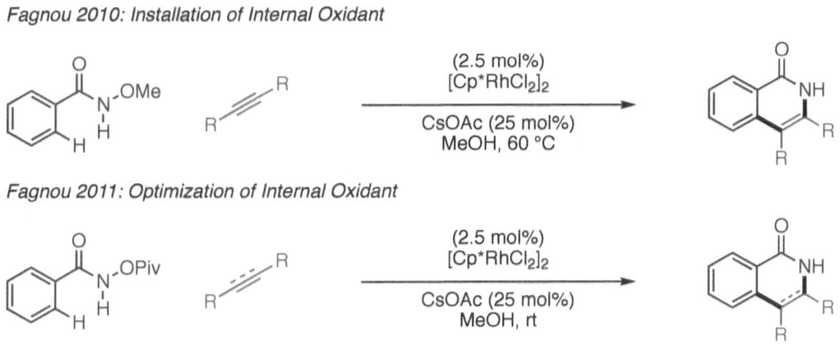

Figure 1.6 Installation of internal oxidant for the oxidative cyclization of benzamides with alkenes and alkynes.

1.4 Mechanistic Considerations

Mechanistically, this system is proposed to work by N–H deprotonation, CMD, and migratory insertion. Reductive elimination of the C–N bond is followed by oxidative addition of the N–O bond. Finally, protodemetallation furnishes the product and regenerates the catalyst.

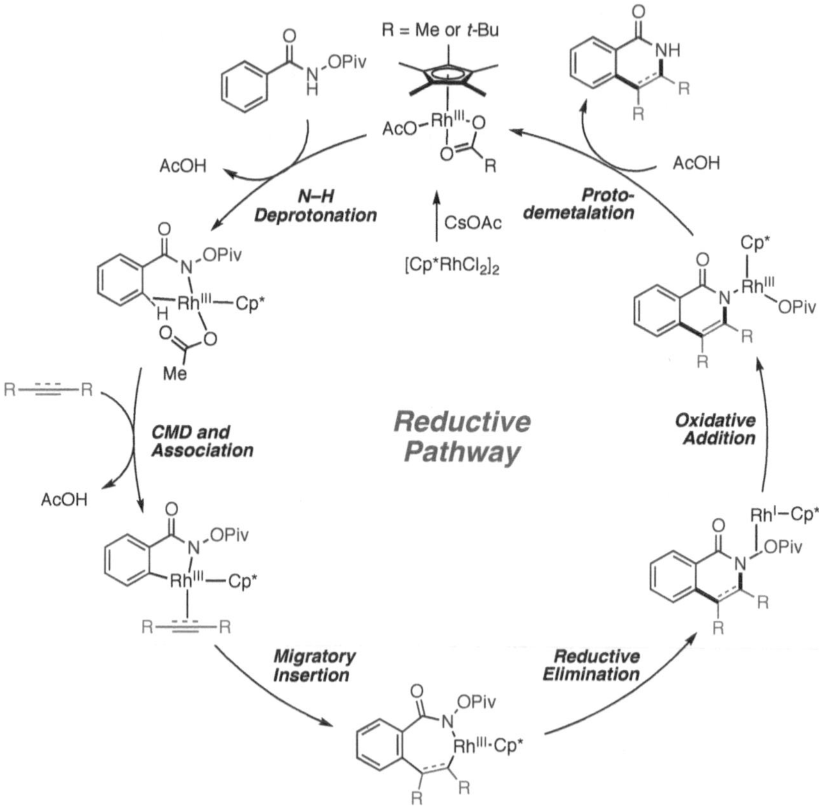

Figure 1.7 *Proposed mechanism of alkene/alkyne benzannulation with an internal oxidant.*

Fagnou's addition of the internal oxidant was revolutionary to Rh(III)-catalyzed C–H activation and alkene difunctionalization. While this proposed mechanism is perfectly reasonable, computational studies and related reactions have given new insights.[7] The major difference comes from the C–N bond forming event being reductive or oxidative in nature.

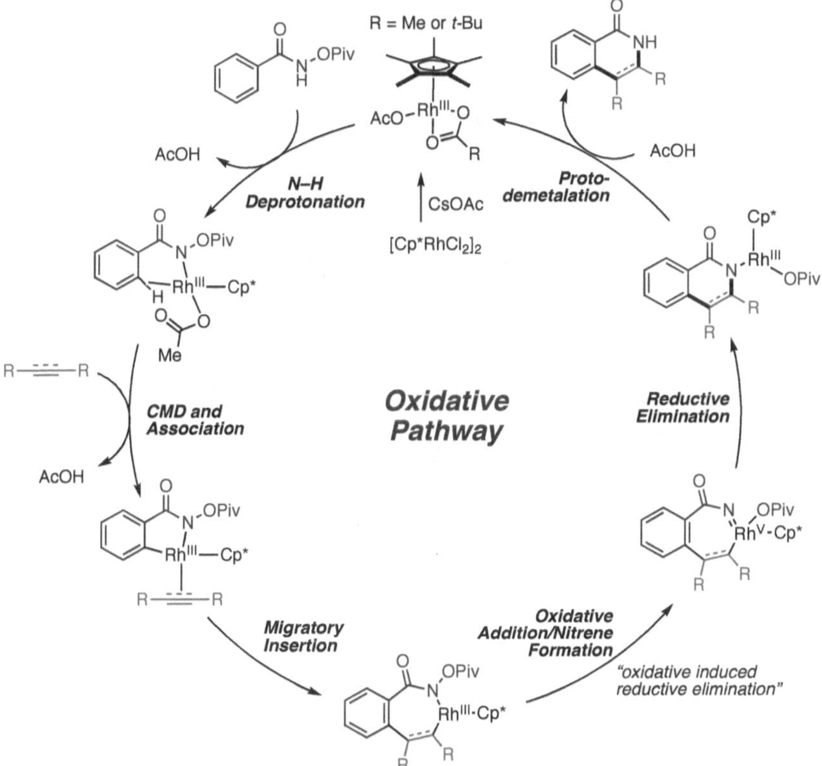

Figure 1.8 *Proposed oxidative nitrene formation.*

In the oxidative pathway, a metal-nitrene is formed by formal oxidation of the Rh center. The idea of *oxidative induced reductive elimination* has gained popularity in recent years among transition metal-catalyzed reactions.[8] In this process, the nitrogen of the benzamide takes on *electrophilic character*. These ideas have had an impact on my own research as well as the field, as seen in the chapters to come.

1.5 Tuning Cyclopentadienyl Ligands to Impact Catalysis

Fundamental studies in the field of Rh(III)-catalyzed C–H activation have deployed Cp* as the parent cyclopentadienyl ligand. In metal-catalyzed reactions, the choice of ligand on the metal affects each step in the catalytic cycle, influencing reactivity and/or selectivity. In the past decade, our group and others have concocted a library of Cp ligands with varying electronic and steric properties on Rh complexes (figure 1.9).[9] Employing these modified Cp ligands as pre-catalysts has affected the reactivity and selectivity the catalysts show in the synthesis of small molecules.[10]

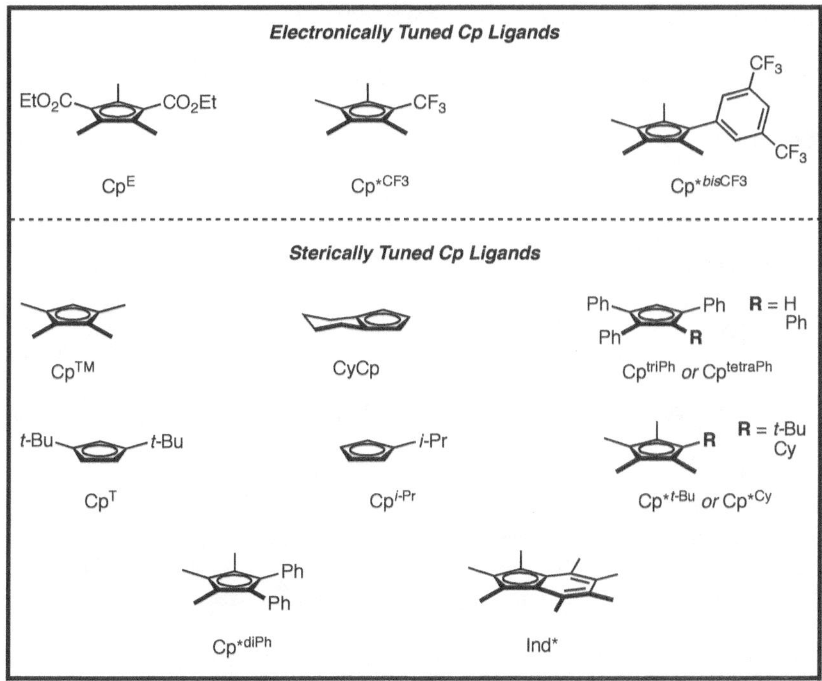

Figure 1.9 *Selected examples of modified cyclopentadienyl ligands.*

1.6 Summary

The direct conversion of carbon–hydrogen bonds into valuable carbon-carbon and carbon-heteroatom bonds is a significant challenge to synthetic organic chemists. More than ever, chemists are employing Rh(III)-catalysts bearing cyclopentadienyl (Cp) ligands to transform otherwise inert C–H bonds. Furthermore, manipulating the sterics and electronics of the Cp ligand has significant impact on catalytic transformations. Our group and others have developed a library of Cp^{X}Rh(III)-precatalysts in hopes of enhancing known reactivity as well as discovering new C–H bond functionalizations.

1.7 References

(1)   a) Colby, D. A.; Bergman, R. G.; Ellman, J. A. *Chem. Rev.* **2010**, *110*, 624. b) Satoh, T.; Miura, M. *Chem. Eur. J.* **2010**, *16*, 11212. c) Patureau, F. W.; Wencel-Delord, J.; Glorius, F. *Aldrichim. Acta* **2012**, *45*, 31. d) Song, G.; Wang, F.; Li, X. *Chem. Soc. Rev.* **2012**, *41*, 3651.

(2)   Lapointe, D.; Fagnou, K. *Chem. Lett.* **2010**, *39*, 1118.

(3)   Walsh, A. P.; Jones, W. D. *Organometallics* **2015**, *34*, 3400.

(4)   Hyster, T. K.; Rovis, T. *J. Am. Chem. Soc.* **2010**, *132*, 10565.

(5)   Guimond, N.; Gouliaras, C.; Fagnou, K. *J. Am. Chem. Soc.* **2010**, *132*, 6908.

(6)   Guimond, N.; Gorelsky, S. I.; Fagnou, K. *J. Am. Chem. Soc.* **2011**, *133*, 6449

(7)   a) Yang, Y.-F.; Houk, K. N.; Wu, Y.-D. *J. Am. Chem. Soc.* **2016**, *138*, 6861. b) Vásquez-Céspedes, S.; Wang, X., Glorius, F. *ACS Catal.* **2018**, *8*, 242.

(8)   a) Bour, J. R.; Camasso, N. M.; Sanford, M. S. *J. Am. Chem. Soc.* **2015**,

137, 8034. b) Kim, J.; Shin, K.; Jin, S.; Kim, D.; Chang, S. *J. Am. Chem. Soc.* **2019**, *141*, 4137. c) Harris, R. J.; Park, J.; Nelson, T. F.; Iqbal, N.; Salgueiro, D. C.; Basca, J.; MacBeth, C. E.; Baik, M.-H.; Blakey, S. *J. Am. Chem. Soc.* **2020**, *142*, 5842.

(9)   a) Piou, T.; Rovis, T. *Acc. Chem. Res.* **2018**, *51*, 170. b) Romanov Michaelidis, F.; Phipps, E. J. T.; Rovis, T. *Chapter 20 of Rhodium Catalysis in Organic Synthesis: Methods and Reactions.* **2019**, 593.

(10)   Piou, T.; Romanov-Michailidis, F.; Romanova-Michaelides, M.; Jackson, K. E.; Semakul, N.; Taggart, T. D.; Newell, B. S.; Rithner, C. D.; Paton, R. S.; Rovis, T. *J. Am. Chem. Soc.* **2017**, *139*, 1296.

– Chapter Two –

Rh(III)-catalyzed Cyclopropanation of
Unactivated Alkenes Initiated by C–H Activation

2.1 Introduction to Cyclopropanation

The synthesis of cyclopropane-containing molecules has intrigued synthetic organic chemists for years because of their prevalence in synthetic targets[1] as well as their susceptibility as reactive intermediates.[2]

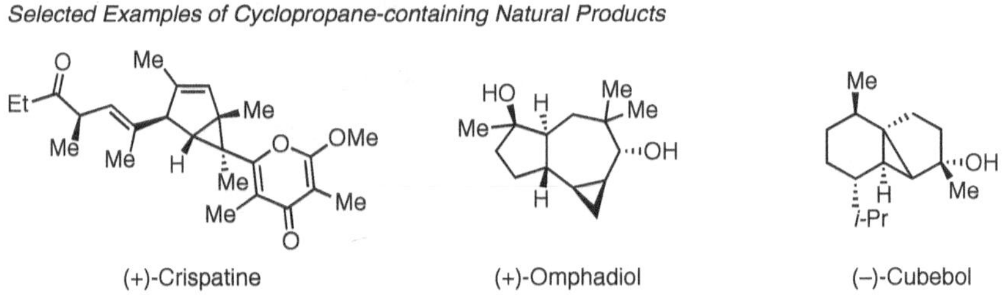

Selected Examples of Cyclopropane-containing Natural Products

(+)-Crispatine (+)-Omphadiol (–)-Cubebol

Figure 2.1 *Selected examples of cyclopropane units in natural product synthesis.*

Preferably, cyclopropane ring construction would be an intermolecular reaction involving a 1-carbon unit adding to a 2-carbon unit which is formally a [2+1] annulation. The simplest 2-carbon units for use in the synthesis are alkenes. Generally, when probing for new reactivity using an alkene, chemists tend to start with activated alkenes bearing an electron withdrawing or donating group to help polarize the alkene. Alkenes bearing only alkyl groups have and continue to remain a challenge due to the chemical inertness.

Furthermore, starting alkene geometry can translate to stereodefined cyclopropane products.

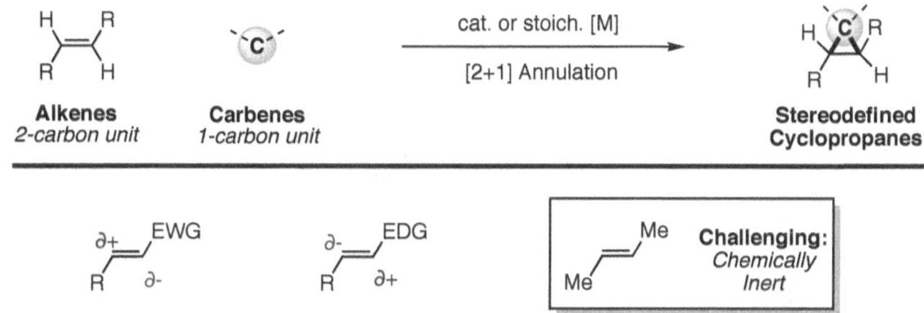

Figure 2.2 General protocol for the synthesis of cyclopropanes from alkenes.

Regarding the 1-carbon units, carbene precursors are known to be effective due to

A plethora of robust methods have been developed to afford cyclopropane motifs from alkenes. Generally, Simmons-Smith and diazo decomposition are regarded as the two most powerful methods for the cyclopropanation of alkenes.[3, 4] Simmons-Smith type reactions are well-established to afford cyclopropanes from the generation of a zinc-carbenoid species that interacts with unactivated olefins with high stereoselectivity; however, these methods are limited by the substitution pattern of the carbenoid reagent[5] and the stoichiometric use of zinc.[6] Regarding unactivated olefins, Uyeda and coworkers noted that under standard Simmons-Smith type conditions, cyclopropanation of non-conjugated dienes is moderate in yield and only moderately selective for the terminal alkene.[7] However, the addition of a Co catalyst bearing a pyridyldiimine ligand is able to distinguish between the two alkenes and perform in good yield. Uyeda demonstrated the power of this method on a number of similar unactivated alkenes.

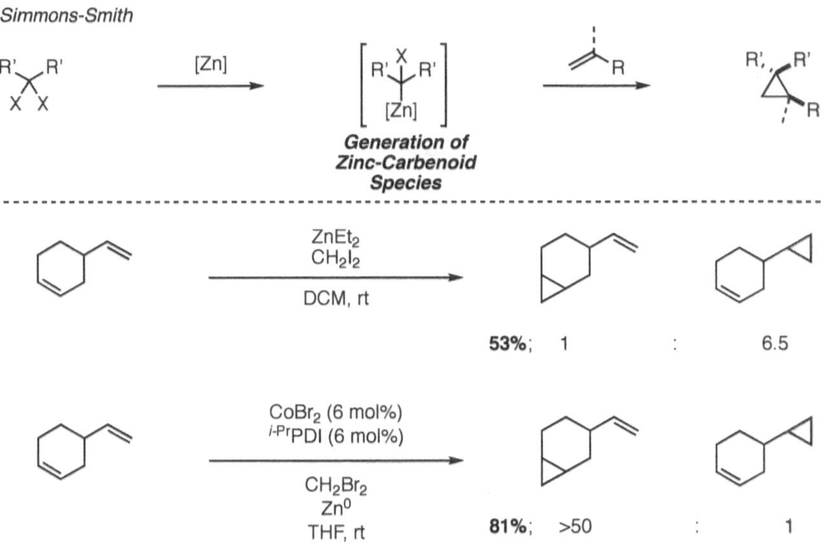

Figure 2.3 *Simmons-Smith reactivity with unactivated alkenes.*

Metal-catalyzed diazo decomposition has also provided complimentary reactivity to access stereodefined cyclopropanes with a more diverse substitution pattern albeit with two notable shortcomings. While numerous methods have been established for Rh-catalyzed cyclopropanation of alkenes,[8] many of these methods require the use of high-energy diazo compounds.[9] Davies and coworkers have been arguably the biggest influence on this chemistry for decades. Cyclopropanation of unactivated alkenes using Rh(II) catalysts has been well known and demonstrated to work with a variety of alkenes.

Notably, Davies and coworkers employed a Rh(II) catalyst bearing protected proline ligand to impart enantioselectivity on the transformation.[10] These Rh(II) catalysts with other chiral ligands have cemented themselves to the field of metal-carbene transfer chemistry. Interestingly, to date, Fürstner and coworkers published the only example Rh(III)-catalyzed cyclopropanation of styrene type alkenes from diazo one-carbon components.

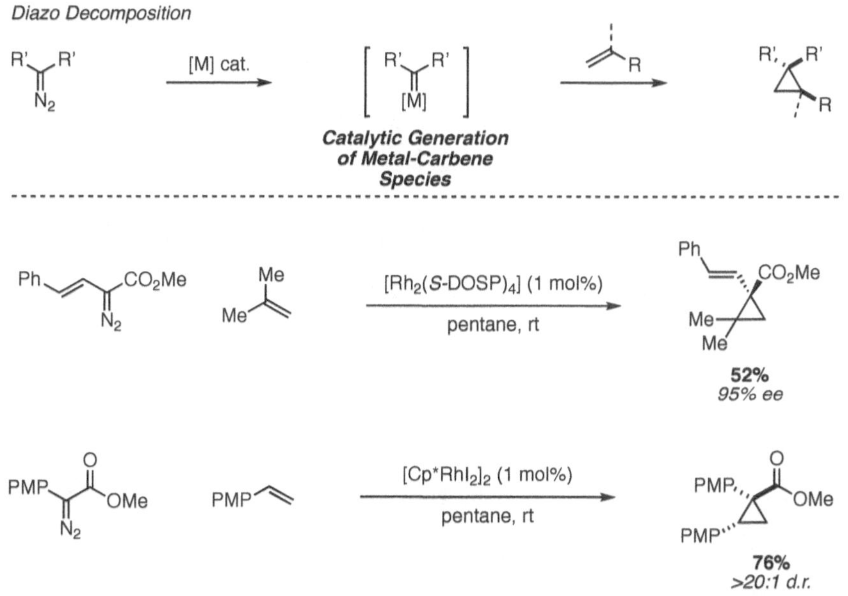

Figure 2.4 Metal catalyzed diazo decomposition of unactivated alkenes.

2.2 Reactivity Profile of *N*-enoxyimides

N-Enoxyphthalimides constitute valuable alternatives to potentially explosive diazo compounds and pyrophoric organozinc reagents due to the mild conditions and the allure of C–H functionalization reactions (Figure 1).[11] Our initial report in 2014 showed that aryl *N*-enoxyphthalimides undergo C–H activation and smoothly undergo [2+1] annulation with activated olefins bearing electron withdrawing groups, affording *trans*-cyclopropanes in good yield and diastereoselectivity.[12]

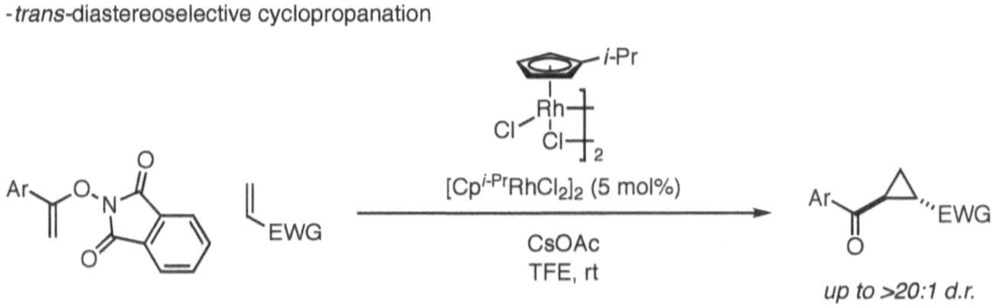

Figure 2.5 Trans-Cyclopropanation.

Importantly, the mechanism first described does not propose the formation of a metal-carbene species. Instead, two migratory insertion events are thought to give rise to the *trans*-cyclopropane. From deuterium labeling studies scrambling is observed alpha to the ketone, indicating a reversible event during the catalytic cycle. To account for this, beta-hydride elimination is proposed to be reversible by Rh–H deprotonation. Due to the high pKa measured of Cp*Rh–H species,[13] this event is unlikely with acetate base as the most viable candidate.

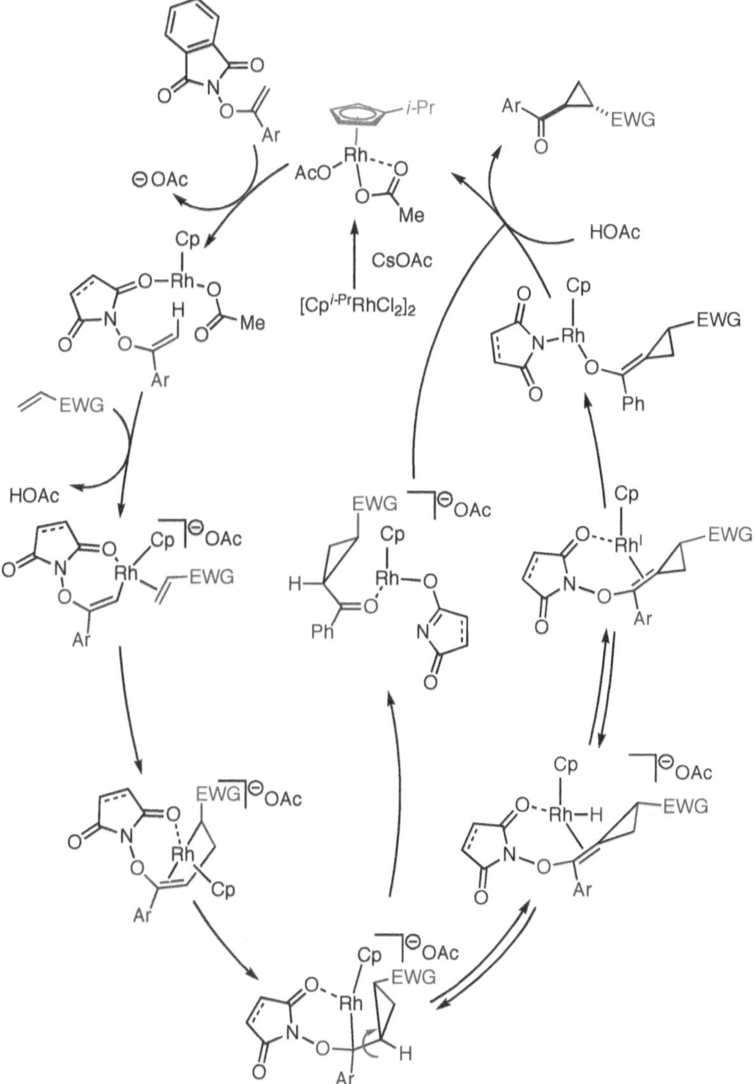

Figure 2.6 *Initially proposed mechanism of Rh(III)-catalyzed cyclopropanation.*

In a follow-up report, we demonstrated that tuning the electronic properties of the Cp ligand as well as the phthalimide ring affords access to the *cis*-cyclopropane scaffold.[14]

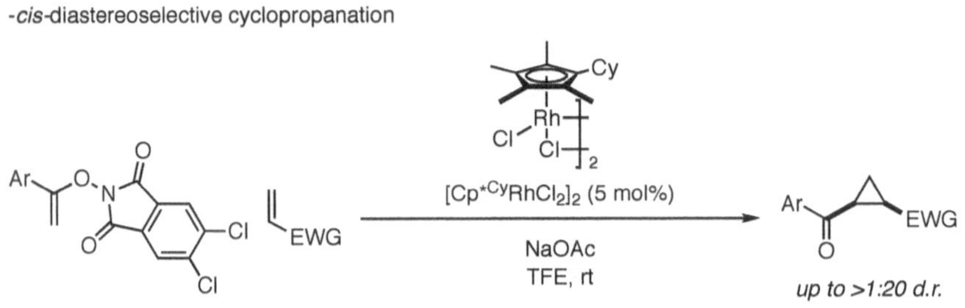

-*cis*-diastereoselective cyclopropanation

Figure 2.7 *Cis-Cyclopropanation.*

Here, the change in selectivity is proposed to arise by phthalimide ring opening by the alcoholic solvent.

Cramer and coworker have rendered the *trans*-cyclopropanation reaction asymmetric by employing their chiral Cp ligand to provide *trans*-cyclopropanes in high e.r.[15] Additionally, they were able to expand the scope of the one-carbon unit beyond aryl substituents.

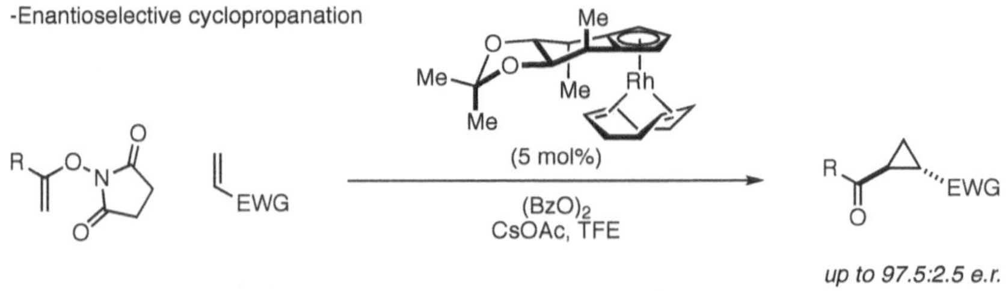

Figure 2.8 *Enantioselective cyclopropanation.*

In an effort to expand the scope of this transformation, we set out to examine stereodefined cyclopropanation of unactivated olefins.

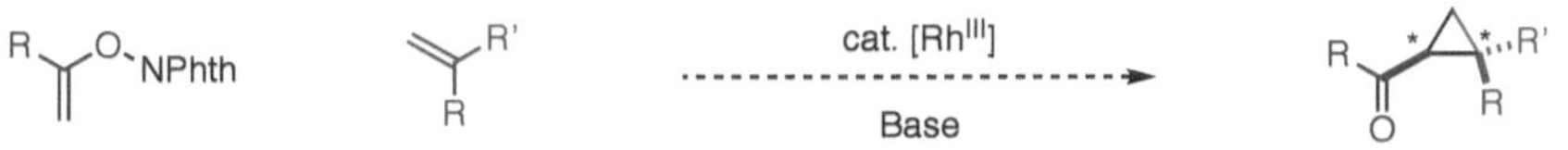

Figure 2.9 Proposed Rh(III)-catalyzed cyclopropanation of unactivated alkenes from N-enoxyphthalimides

2.3 Reaction Optimization

From the *trans*-cyclopropanation study, our group found that 1,1-dialkylalkenes undergo cyclopropanation in modest yield. The shortcoming in this transformation is the reaction is unselective with $[Cp*RhCl_2]_2$ precatalyst.

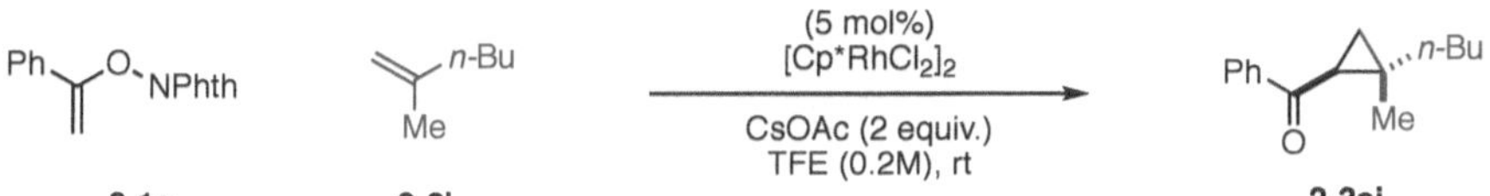

*Figure 2.10 Original hit with Cp*Rh(III) precatalyst.*

Because of the large library of $Cp^XRh(III)$ precatalysts our group has built we predicted that by tuning the sterics and/or electronics of the Cp ligand, we could impart selectivity in cyclopropanation of unactivated alkenes. After screening 15+ ligands, we observed no change in diastereoselectivity. Notably, we found that electron-deficient Cp ligands improve the yield of the reaction. In particular, Cp*^{CF3} gives the highest yield of

82%.

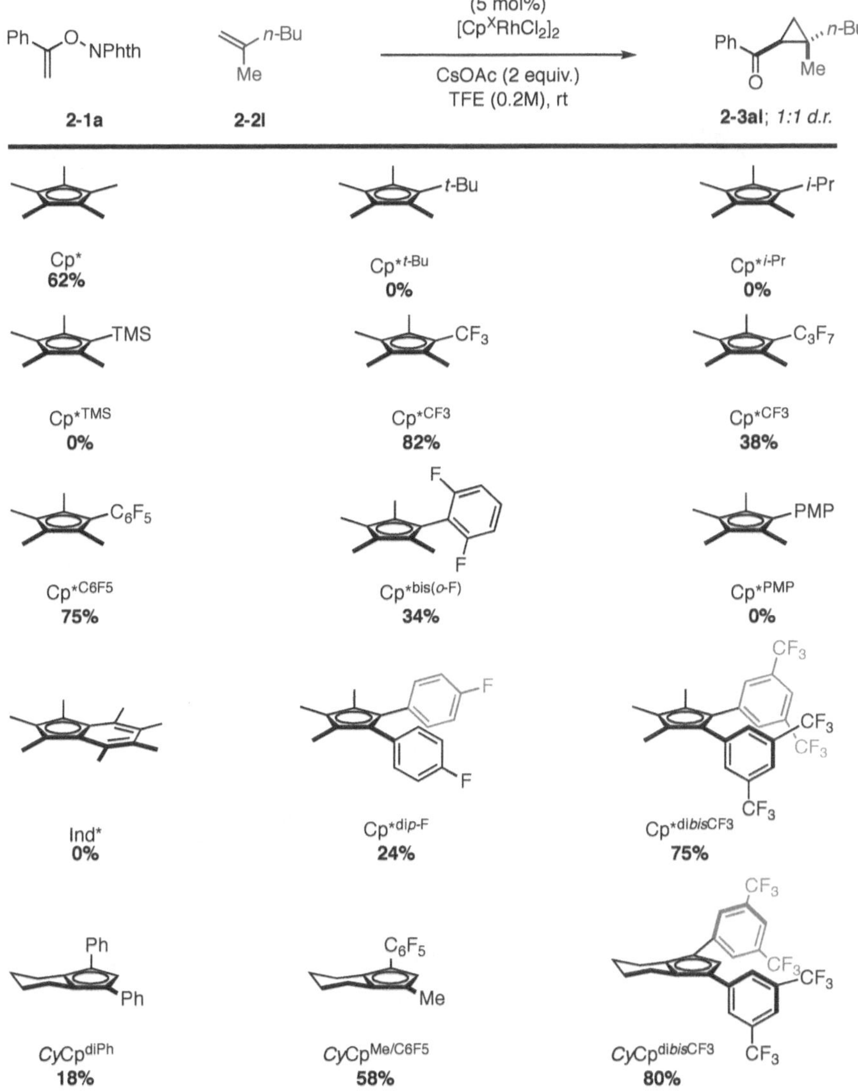

Scheme 2.1 Cp Ligand screen.

Due to the lack of diastereoselectivity imparted by the Cp ligand we chose to advance the project with symmetrical 1,1-disubstituted alkenes in the presence of Cp*^{CF3}Rh(III) catalyst.

2.4 Scope of the Cyclopropanation Reaction

We began by examining 3-methylenepentane as a coupling partner and found modest reactivity as cyclopropane **2-3aa** was afforded in 40% yield. A number of exocyclic alkenes proved to be excellent participants in this reaction giving a wide range of [2.n]spirocyclic ketones. We interrogated the effect of different size carbocycles ranging from 4 to 8-membered rings (**2-3ab** to **2-3af**). Notably, methylenecyclohexane gives [2.5]spirocycle **2-3ad** in near quantitative yield. Both tosyl- and Boc-protected methylene piperidines display good reactivity affording cyclopropane **2-3ag** in 72% and **2-3ah** in 89% yield, respectively. Cyclopropanation of a methylene cyclohexane bearing a substituent at the 4-position proceeds efficiently, delivering cyclopropane **2-3ai** in 97% yield and good diastereoselectivity (8.6:1 d.r.).[16]

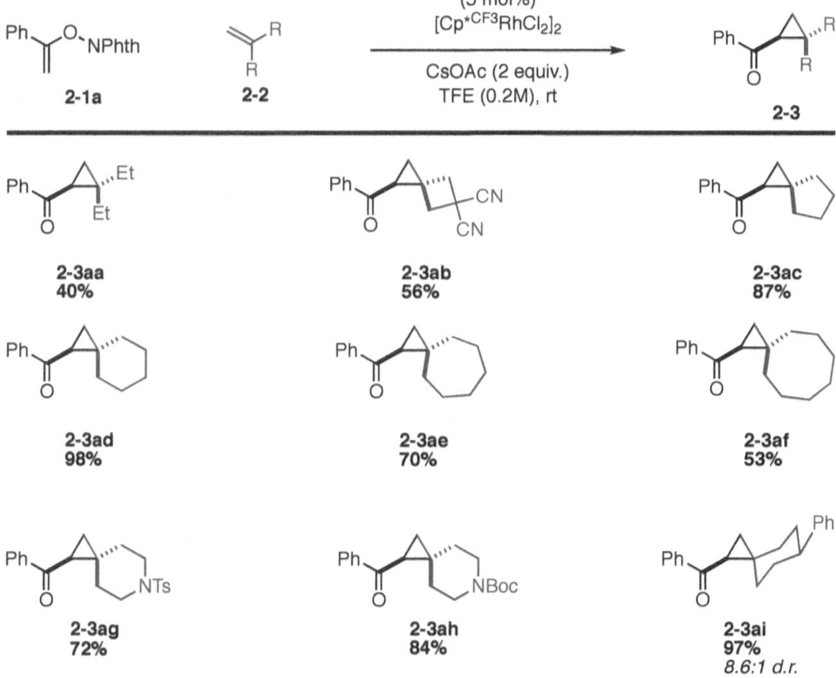

Scheme 2.2 *Scope of 1,1-disubstituted alkenes.*

Varying *para-* (**2-3bd** to **2-3dd**) and *meta-* (**2-3ed** to **2-3gd**) arene substitution on the enoxyphthalimide is tolerated, with each substrate displaying excellent yields. *ortho-*Fluorine containing enoxyphthalimide delivers cyclopropane **2-3hd** in 59% yield. Naphthyl enoxyphthalimide gives cyclopropane **2-3id** in 67% yield. Finally, an alkyl substituted *N*-enoxyphthalimide is also a competent substrate, affording cyclopropane **2-3jd** in 98% yield.

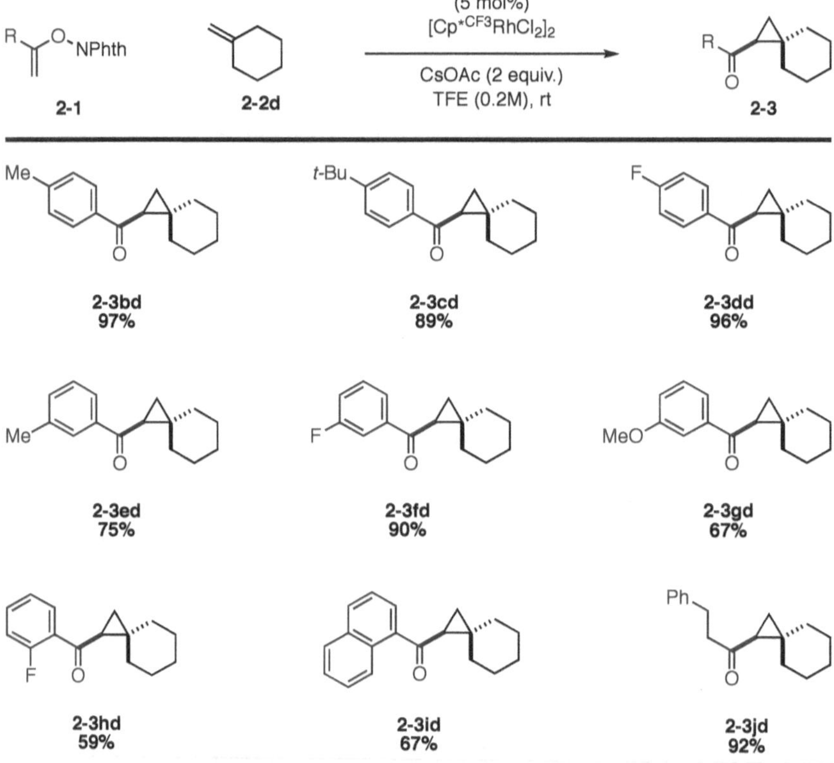

Scheme 2.3 *Scope of N-enoxyphthalimides.*

While these examples display a nice range of functional group tolerance, unactivated alkenes with different substitution patterns behave differently.

2.5 Participation of Other Alkenes

We also surveyed the reactivity pattern of different alkenes: Knowing the cyclopropane **2-3aj** is formed in good yield but unselective, we tried to increase the steric load by changing *n*-Bu to *i*-Pr and saw a dramatic drop in yield, with poor diastereoselectivity. Styrene gives the desired cyclopropane in good yield but poor d.r. Gratifyingly, we see the related unactivated alkene, 1-decene gives cyclopropane **2-3am** in moderate yield as well. Vinyl acetate does not participate in the cyclopropanation reaction; however, the related electron-rich alkene 2,3-dihydrofuran gives cyclopropane **2-3ao** in low yield. Similarly, the locked *cis* alkene cyclopentene and the related *trans*-4-octene proceed in low yield. Interestingly, norbornene provides tricycle **2-3ar** in moderate yield and importantly as a single diastereomer. Combining what we know from the performance of 1,1-disubstituted alkenes and styrenes, we were disappointed that α-methyl styrene does not participate in the cyclopropanation reaction. However, introducing a methylene spacer restores moderate reactivity and 1:1 diastereoselectivity. Finally, similar to substrate **2-3aa**, we see that extending the chain using 5-methylenenonane drops reactivity to 12% yield.

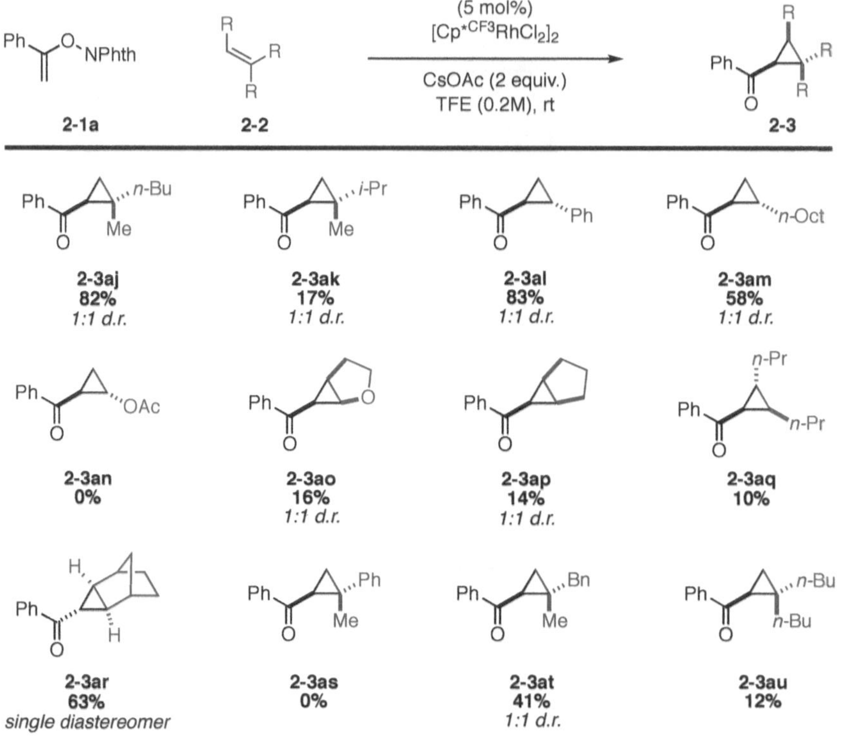

Scheme 2.4 *Scope of alkenes with varying substitution.*

2.6 Mechanistic Studies

Finally, we sought to interrogate the mechanism of this reaction (Figure 4). Subjecting **2-1a** to the reaction conditions using TFE-d_1 leads to no deuterium incorporation upon re-isolation of **2-1a**. In another experiment, we subjected **2-1a** and **2-2d** to the reaction conditions again with TFE-d_1 that gives cyclopropane **2-3ad'** in 85%

yield. From the analysis of the product, we observe a reversible deuterium exchange

event at the alpha-position (54% D incorporation).

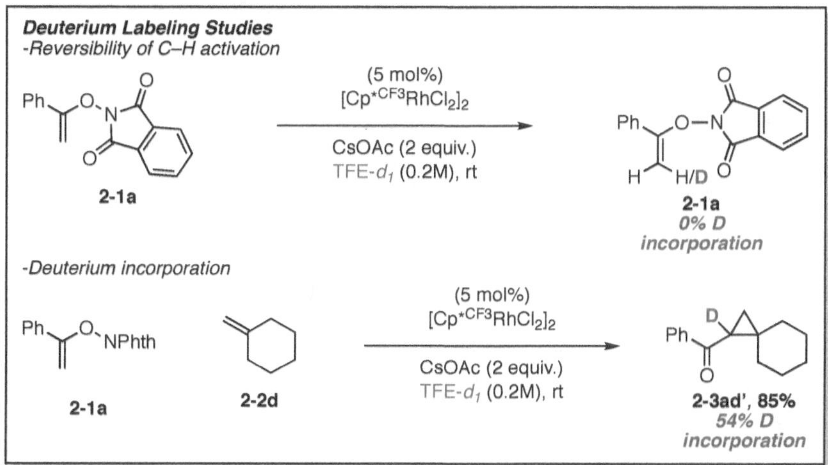

Figure 2.11 Deuterium labeling studies.

We next probed the role of the phthalimide ring by subjecting **2-1a** to 2 equiv. of

CsOAc in TFE and observed the formation of dioxazoline **2-4** in 59% yield, indicating

TFE opens the phthalimide ring. We believe the resulting amide is a key intermediate to

direct the catalyst for the C–H activation step; however, attempts to isolate the opened

phthalimide were unsuccesful. Finally, we subjected **2-4** to **2-2d** and the reaction

conditions. However, only trace product was observed indicating **2-4** does not

significantly contribute as a competent reaction intermediate.

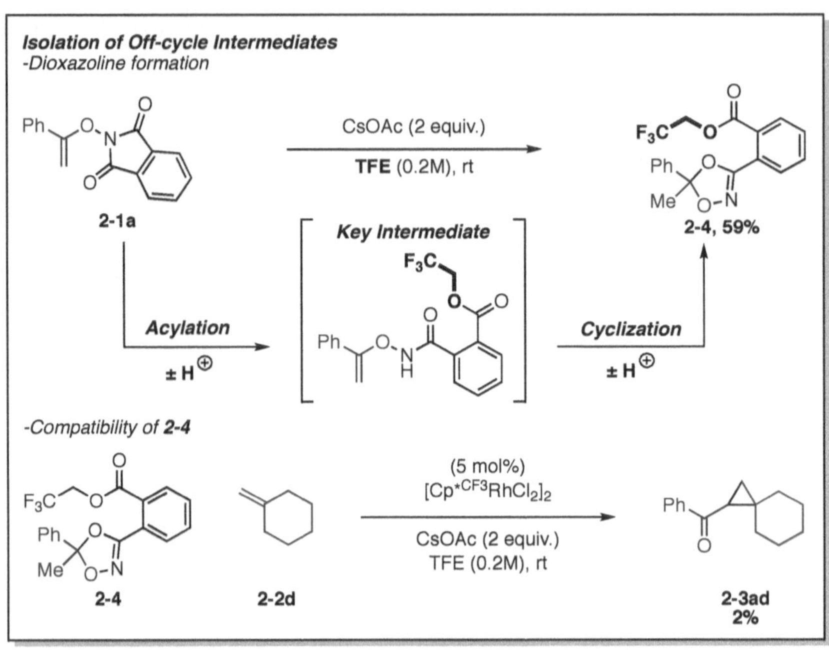

Figure 2.12 Dioxazoline formation and intermediacy test.

2.7 Proposed Mechanism

On the basis of these experiments, we propose the following mechanism:

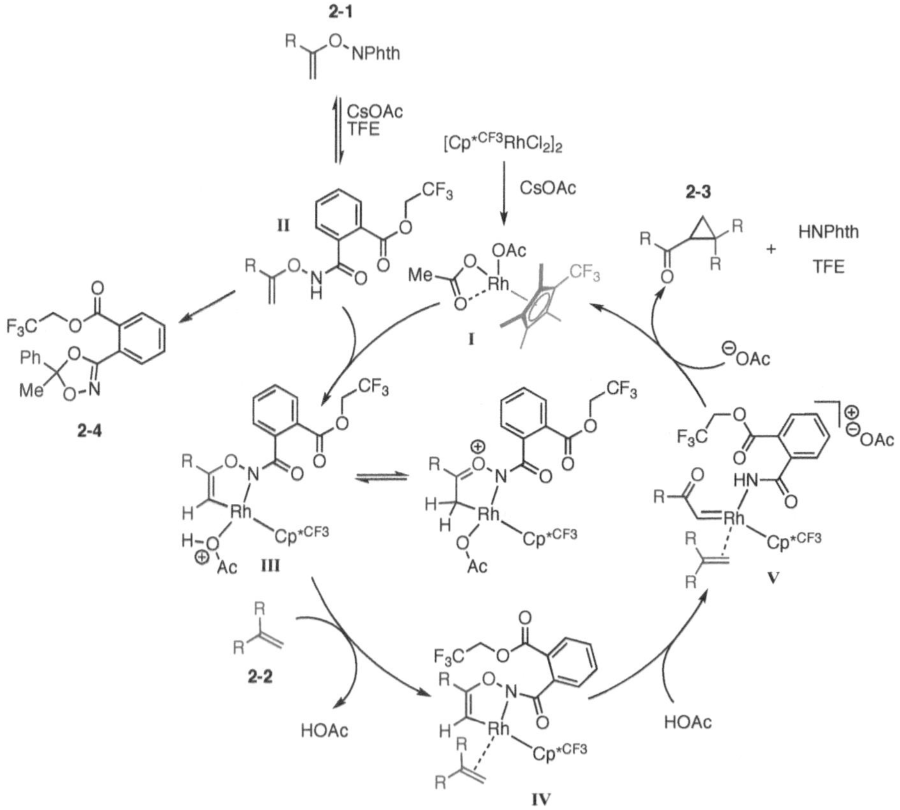

Figure 2.30 Proposed Mechanism.

First, the precatalyst undergoes salt metathesis with CsOAc to form the active

catalyst **I**. Concurrently, **2-1** is opened by the solvent to give **II** which then intercepts **I**, before dioxazoline **2-4** formation, and undergoes C–H activation via concerted metalation-deprotonation to afford intermediate **III**. At this stage, we believe intermediate **III** displays enolic character to reversibly wash in deuterium before ligand exchange of **2-2**. After exchanging acetic acid for alkene that gives intermediate **V**, we propose the formation of a Rh-carbene, intermediate **V**, via cleavage of the N–O bond. Intermediate **V** then gives way to the desired cyclopropane product.

2.8 Summary

In conclusion, we have developed a Rh(III)-catalyzed cyclopropanation protocol for *N*-enoxyphthalimides and unactivated olefins. The *N*-enoxyphthalimide has been shown to undergo C–H activation that leads to a proposed metal-carbene to induce a [2+1] annulation with alkenes that give a diverse range of cyclopropyl ketones in mild conditions.

2.9 References

---

(1) (a) Chen, D.Y.-K.; Pouwer, R. H.; Richard, J.-A. *Chem. Soc. Rev.* **2012**, *41*, 4631. (b) Talele, T. T. *J. Med. Chem.* **2016**, *59*, 8712.

(2) (a) Banwell, M. G.; Edwards, A. J.; Jolliffe, K. A.; Smith, J. A.; Hamel, E.; Verdier-Pinard, P. *Org. Biomol. Chem.* **2003**, *1*, 296. (b) Newhouse, T. R.; Kaib, P. S. J.; Gross, A. W.; Corey, E. J. *Org. Lett.* **2013**, *15*, 1591.

(3) For a recent selection of many diazo decomposition reactions see: (a) Doyle, M. P.; Forbes, D. C. *Chem. Rev.* **1998**, *98*, 911. (b) Davies, H. M. L.; Antoulinakis, E. G. *Org. React.* **2004**, *57*, 1.

(4) For a recent selection of Simmons-Smith-type reactions see: (a) Lebel, H.; Marcoux, J.-F.; Molinaro, C.; Charette, A. B. *Chem. Rev.* **2003**, *103*, 4977. (b) Charette, A. B.; Beauchemin, A. *Org. React.* **2004**, *58*, 1.

(5) (a) Friedrich, E. C.; Biresaw, G. *J. Org. Chem.* **1982**, *47*, 1615. (b) Stahl, K.-J.; Hertzsch, W.; Musso, H.; *Liebigs Ann. Chem.* **1985**, 1474. (c) Roberts, C.; Walton, J. C. *J. Chem. Soc., Perkin Trans. 2* **1985**, 841. (d) Motherwell, W. B.; Roberts, L. R. *Chem. Commun.* **1992**, 1582.

(6) (a) Dolbier Jr., W. R.; Burkholder, C. R. *J. Org. Chem.* **1990**, *55*, 589. (b) Ilchenko, N. O.; Hedberg, M.; Szabó, K. *J. Chem. Sci,* **2017**, *8*, 1056. (c) Werth, J.; Uyeda, C. *Chem. Sci.* **2018**, *9*, 1604.

(7) Werth, J.; Uyeda, C. Chem. Sci. **2018**, *9*, 1604.

(8) For selected recent examples of Rh-catalyzed cyclopropanations see: (a) Muthusamy, S.; Gunanathan, C. *Synlett* **2003**, *11*, 1599. (b) Hilt, G.; Galbiati, F. *Synthesis,* **2006**, *21,* 3589. (c) Lindsay, V. N. G.; Lin, W.; Charette, A. B. *J. Am. Chem. Soc.* **2009**, *131*, 16383. (d) Lindsay, V. N. G.; Nicolas, C.; Charette, A. B. *J. Am. Chem. Soc.* **2011**, *133*, 8972. (e) Negretti, S.; Cohen, C. M.; Chang, J. J.; Guptill, G. M.; Davies, H. M. L. *Tetrahedron* **2015**, *71*, 7415. (f) Lehner, V.; Davies, H. M. L.; Reiser, O. *Org. Lett.* **2017**, *19*, 4722. (g) Sun, G.-J.; Gong, J.; Kang, Q. *J. Org. Chem.* **2017**, *82*, 1796. (h) Tindall, D. J.; Werle, C.; Goddard, R.; Philipps, P.; Fares, C.; Fürstner, A. *J. Am. Chem. Soc.* **2018**, *140*, 1884. (i) Lindsay, V. N G. Rhodium(II)-Catalyzed Cyclopropanation. In *Rhodium Catalysis in Organic Synthesis: Methods and Reactions*; Tanaka, K., Ed.; Wiley-VCH; 2018; pp. 433-448.

(9) (a) Doyle, M. P.; Hu, W.; Phillips, I. M.; Moody, C. J.; Pepper, A. G.; Slawin, A. M. *Adv. Synth. Catal.* **2001**, *343*, 112. (b) Doyle, M. P.; Hu, W. *Adv. Synth. Catal.* **2001**, *343*, 299. (c) Gharpure, S. J.; Shukla, M. K.; Vijayasree, U. *Org. Lett.* **2009**, *11*, 5466. (d) Vanier, S. F.; Larouche, G. Wurz, R. P.; Charette, A. B. *Org. Lett.* **2009**, *12*, 672. (e) Nani, R. R.; Reisman, S. E. *J. Am. Chem. Soc.* **2013**, *135*, 7304. (f) Gu, H.; Huang, S.; Lin, X. *Org. Biomol. Chem.* **2019**, *17*, 1154.

(10) Davies, H. M. L.; Bruzinski, P. R.; Lake, D. H.; Kong, N.; Fall, M. J. *J. Am. Chem. Soc.* **1996**, *118*, 6897.

(11) (a) Doyle, M. P.; Duffy, R.; Ratnikov, M.; Zhou, L. *Chem. Rev.* **2010**, *110*, 2704. (b) Colby, D. A.; Tsai, A. S.; Bergman, R. G.; Ellman, J. A. *Acc. Chem. Res.* **2012**, *45*, 6814. (c) Piou, T.; Rovis, T. *Acc. Chem. Res.* **2018**, *51*, 1170.

(12) Piou, T.; Rovis, T. *J. Am. Chem. Soc.* **2014**, *136*, 11292.

(13) Hu, Y.; Norton, J. R. *J. Am. Chem. Soc.* **2014**, *136*, 5938.

(14) Piou, T.; Romanov-Michailidis, F.; Ashley, M. A.; Romanova- Michaelides, M.; Rovis, T. *J. Am. Chem. Soc.* **2018**, *140*, 9587.

(15) Duchemin, C.; Cramer, N. *Chem. Sci.* **2019**, *10*, 2773.

(16) Diastereoselectivity assigned by analogy of other 3-membered rings formed from 4-substituted-exocyclic alkenes: (a) Corey, E. J.; Chaykovsky, M. *J. Am. Chem. Soc.* **1965**, *87*, 1353. (b) Carlson, R. G.; Behn, N. S. *J. Org. Chem.* **1967**, *32*, 1363. (c) Bellucci, G.; Chiappe, C.; Lo Moro G.; Ingrosso, G. *J. Org. Chem.* **1995**, *60*, 6214.

– Chapter Three –

Rh(III)-catalyzed C–H Activation-Initiated
Directed Cyclopropanation of Allylic Alcohols

3.1 Cyclopropanation of Allylic Alcohols

Biological and synthetic targets containing cyclopropane units have intrigued organic chemists for years as a result of their unique properties and the synthetic challenges.[1] A number of powerful methods have been developed for the stereoselective synthesis of cyclopropane motifs.[2] These methods largely share a common approach of an alkene that undergoes a [2+1] annulation with carbenes, metal carbenes, or metal-carbenoid species. In particular, allylic alcohols have been exploited as coupling partners in cyclopropanation reactions for their leverageable, pendent hydroxyl group. Ultimately, this handle provides regio- and diastereoselective cyclopropanations. Two methods have emerged as preferred techniques for the cyclopropanation of alkenes: Simmons-Smith type reactions and catalyzed diazo decompositions.

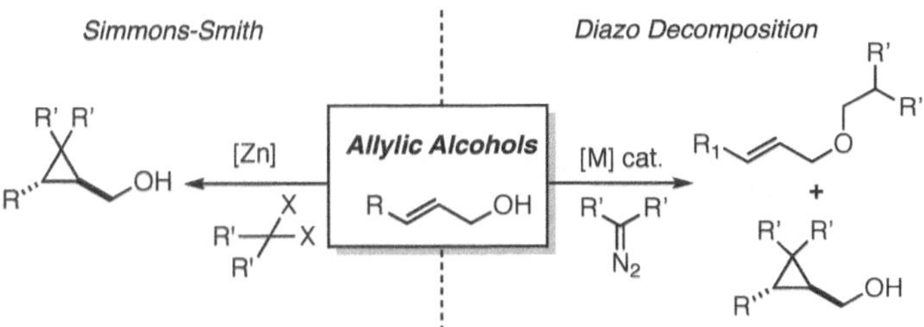

Figure 3.1 *General strategies for the cyclopropanation of allylic alcohols.*

The Simmons-Smith approach features stoichiometric zinc reagents to aid both the formation and transfer of carbenoid species from simple methylene sources. Similarly, metal-catalyzed diazo decomposition is a broadly powerful reactivity manifold for the cyclopropanation of alkenes, with Rh,[3] Ru,[4] Pd,[5] Cu,[6] Co,[7] and Fe[8] catalysts utilized for their carbenoid formation and transfer capabilities. Notably, both modes of reactivity have also been rendered asymmetric when using prochiral alkenes.[9] Charette has implemented strategies for enantioselective cyclopropanation of unprotected allylic alcohols by employing chiral diamine ligands or catalytic Ti bearing a taddolate ligand for chirality transfer. Additionally, in the realm of asymmetric cyclopropanation, metal catalysts (Cu and Rh shown below) bearing chiral ligands have been used to decompose diazo compounds and undergo [2+1] annulation with protected allylic alcohols in stereoselective fashion. Here it is necessary for the allylic alcohol to be protected to minimize unwanted byproducts.

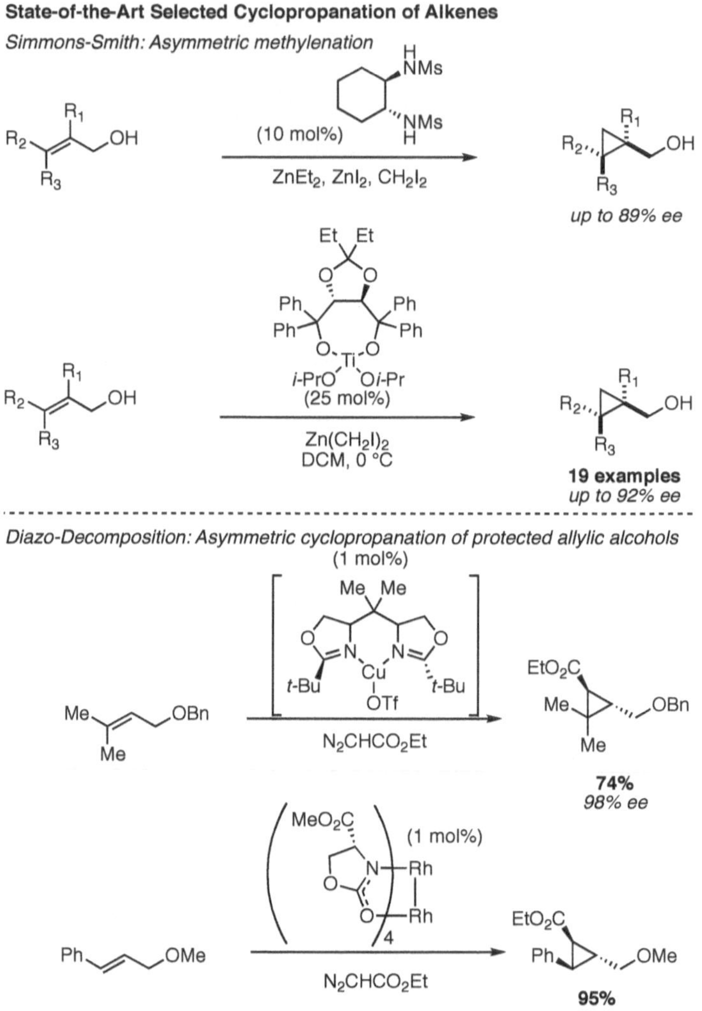

Figure 3.2 State-of-the-Art strategies for cyclopropanation of allylic alcohol-type alkenes.

With regards to allylic alcohols, notable shortcomings have arisen in the two

established methods outlined above. While Simmons-Smith reactivity is regio-, and diasteroselective, it is largely limited to methylenation[10] Charette has shown that pre-functionalizing substituted methylene-zinc precursors allows for some functional groups to be carried through the cyclopropanation reaction. However, the substitution pattern of the one-carbon unit is limited to iodo- and boryl-functionalized units. In the case of metal-catalyzed diazo decomposition, the cyclopropanation of allyl alcohol is low yielding and instead O–H insertion is observed as the major product. Because the metal-carbene species is electrophilic, the pendant hydroxyl group reacts much faster than the alkene.

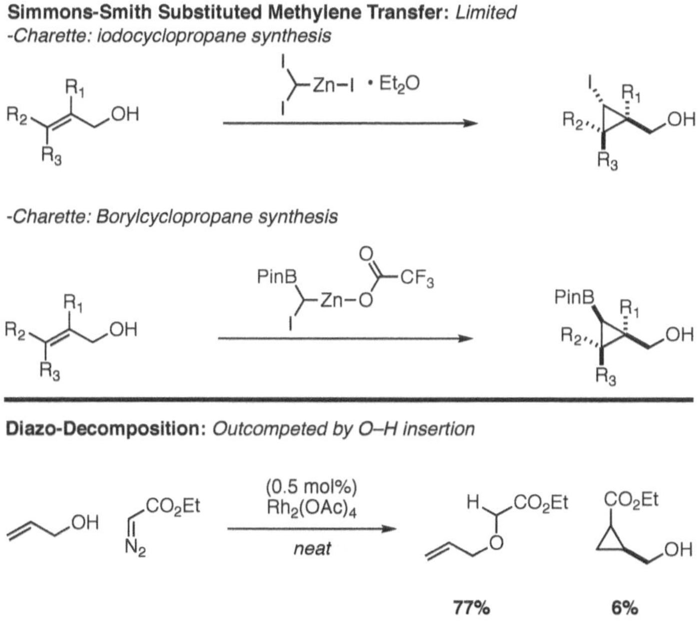

Figure 3.3 Limitations of competitive cyclopropanation strategies.

We have previously reported that *N*-enoxyphthalimides are a unique one-carbon component for the cyclopropanation of activated alkenes.[11] Furthermore, tuning the cyclopentadienyl (Cp) ligand on the Rh[III] catalyst delivers either *cis*- or *trans*-disubstituted cyclopropanes stereoselectively.[12, 13] In a complementary approach, we found that by exchanging trifluoroethanol (TFE) solvent for methanol (MeOH) and again tuning the Cp ligand on the Rh catalyst, activated alkenes undergo *syn*-1,2-carboamination.[14]

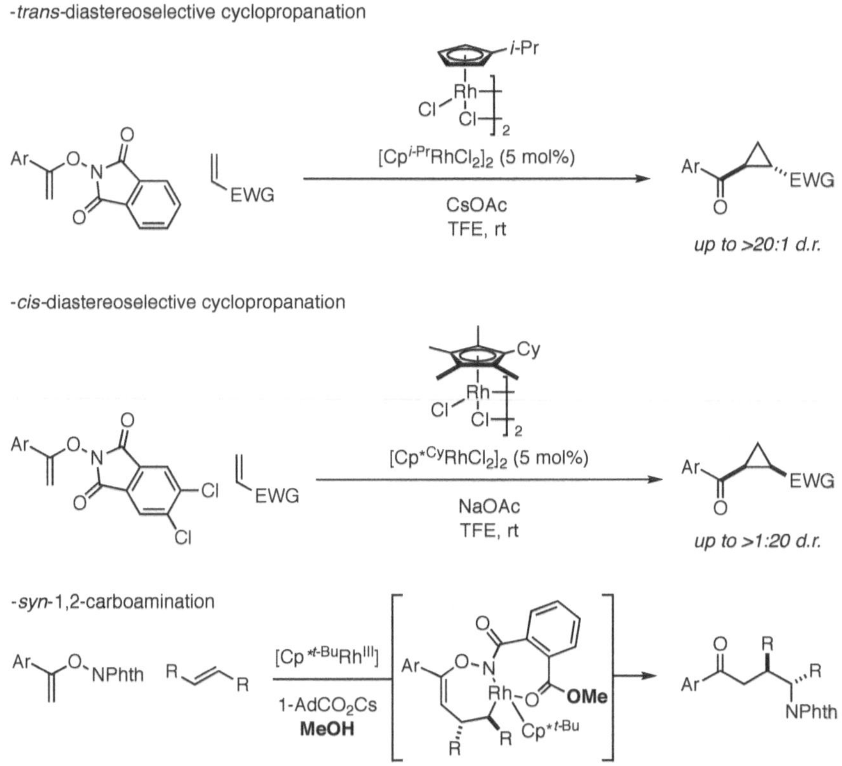

Figure 3.4 Previously described transformations with N-enoxyphthalimides.

This chemodivergence is hypothesized to originate from MeOH participating as a nucleophile to open the phthalimide ring that allows the *N*-enoxyphthalimide to act as a bidentate ligand throughout catalysis. On the basis of these findings, we sought to expand the scope of our reported diastereoselective cyclopropanation toward unactivated alkenes.

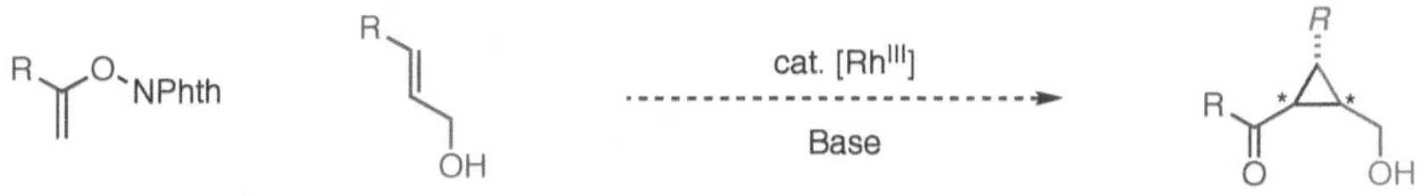

Figure 3.5 Proposed Rh(III)-catalyzed directed cyclopropanation of allylic alcohols.

3.2 Reaction Optimization

Initial investigations began with phenyl-*N*-enoxyphthalimide **3-1a** and *trans*-2-hexen-1-ol **3-2a** in the presence of various Rh(III) catalysts in TFE at room temperature delivering cyclopropane **3aa** in moderate yield but high diastereoselectivities. Ultimately, electron-deficient ligands proved best for this transformation–with Cp*[CF3]

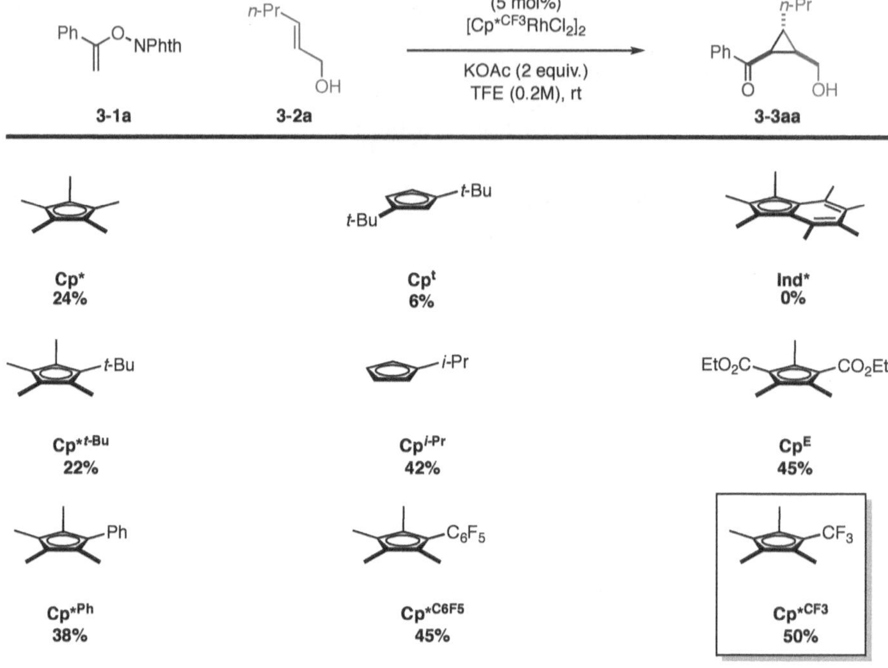

Figure 3.6 Cp ligand optimization.

Solvent (entries B and C) and base screens revealed that KOPiv in TFE is optimal, providing 64% yield and >20:1 *d.r.* for the desired product (entry D). Our next thought was to heat the reaction to push it to completion; however, we observed only 11% yield of product. Furthermore, we discovered that reducing the reaction temperature to 0 °C leads to the desired cyclopropane in 81% yield while preserving excellent diastereoselectivity (entry F).

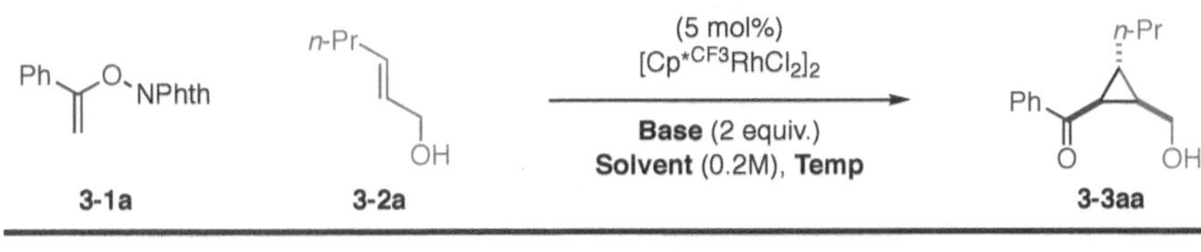

Entry	Base	Solvent	Temperature	Yield
A	KOAc	TFE	rt	50%
B	KOAc	MeOH	rt	36%
C	KOAc	THF	rt	29%
D	KOPiv	TFE	rt	64%
E	KOPiv	TFE	60 °C	11%
F	KOPiv	TFE	0 °C	81%

Scheme 3.1 Reaction optimization–Examination of the effects of inorganic bases, solvents, and temperature.

3.3 Stereoselectivity of the Cyclopropanation Reaction

We next examined if the diastereoselectivity of the tri-substituted cyclopropane product was directly correlated with initial alkene geometry (Scheme 2). Both *trans-* and *cis*-1,2-disubstituted primary allylic alcohols provide the desired cyclopropanes **3-3aa**

and **3-3ab** in good yield–81% and 62%, respectively–and >20:1 *d.r.,* implicating a
stereospecific transformation.

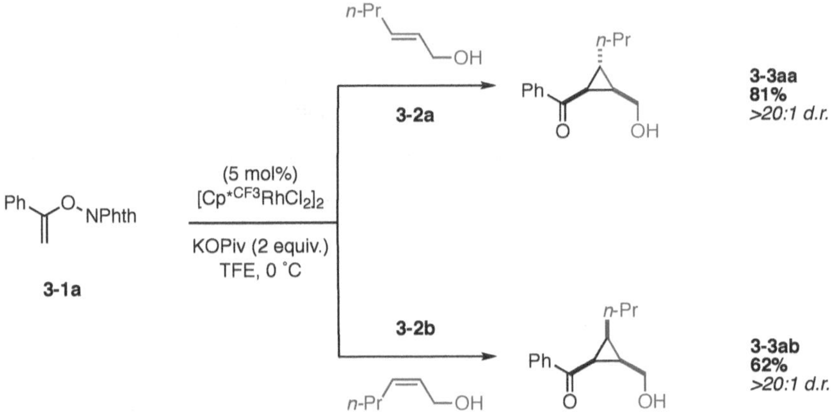

Figure 3.7 Primary allylic alcohols bearing a trans or cis disubstituted alkene.

3.4 Scope of the Cyclopropanation Reaction

Similar to the parent allylic alcohol, we found crotyl alcohol gives cyclopropane
3-3ac in excellent diastereoselectivity and 81% yield. Methallyl alcohol gives
cyclopropane **3-3ad** in 62% yield with 7:1 *d.r.* while prenyl alcohol furnishes **3-3ae** in
82% yield and >20:1 *d.r.*

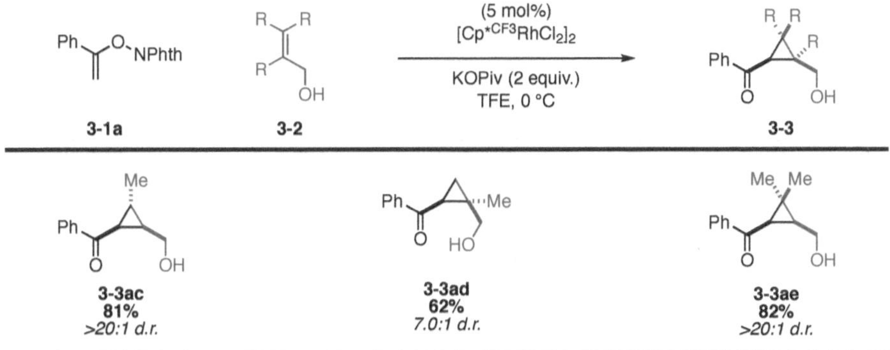

Scheme 3.2 Scope of primary allylic alcohols.

With optimized conditions in hand, we examined the scope of this reaction (Scheme 3). Varying *para*- (**3-3ba-3-3ea**) and *meta*- (**3-3fa-3-3ha**) arene substitution on the enoxyphthalimide is tolerated, with each substrate displaying >20:1 diastereoselectivity. *Ortho*-Fluorine containing enoxyphthalimide delivers cyclopropane **3-3ia** in 44% yield. Alkyl substituted *N*-enoxyphthalimide[15] is also a competent substrate, affording cyclopropane **3-3ka** in 92% yield.

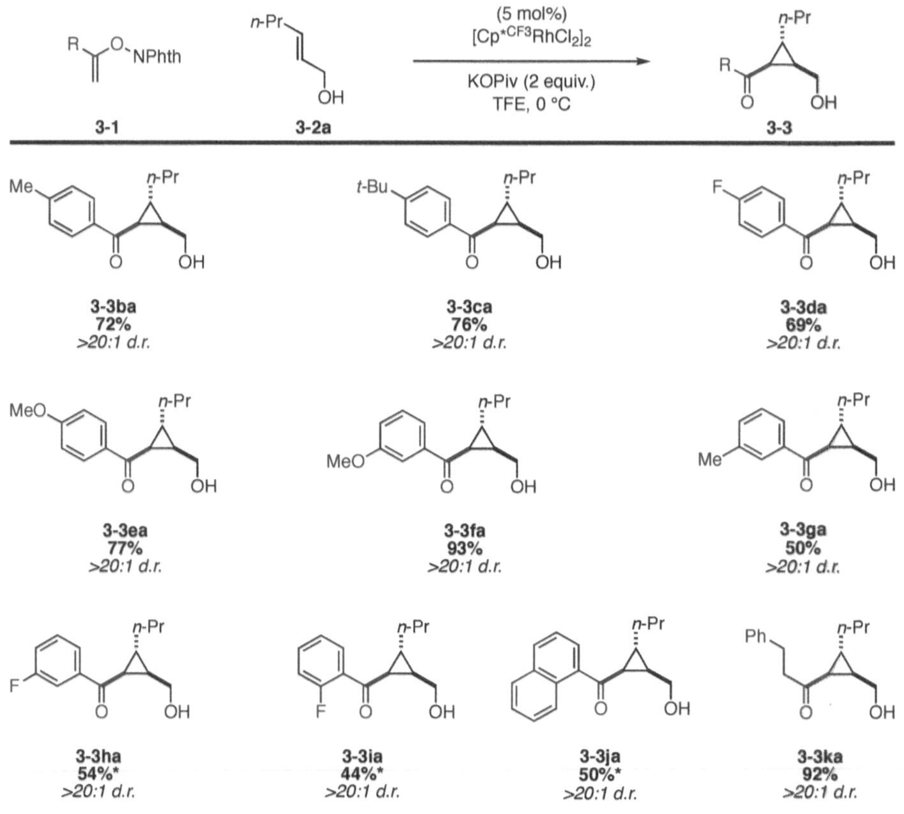

3-1   3-2a   (5 mol%) [Cp*CF3RhCl2]2 / KOPiv (2 equiv.) TFE, 0 °C   3-3

3-3ba
72%
>20:1 d.r.

3-3ca
76%
>20:1 d.r.

3-3da
69%
>20:1 d.r.

3-3ea
77%
>20:1 d.r.

3-3fa
93%
>20:1 d.r.

3-3ga
50%
>20:1 d.r.

3-3ha
54%*
>20:1 d.r.

3-3ia
44%*
>20:1 d.r.

3-3ja
50%*
>20:1 d.r.

3-3ka
92%
>20:1 d.r.

*Low conversion at 0 °C, isolated yield at 21 °C

Scheme 3.3 *Scope of N-enoxyphthalimides.*

Next, a range of suitable allylic alcohols for the cyclopropanation reaction was explored (Scheme 4). Notably, chiral allylic alcohol substrates provide additional complexity leading to the potential of four different stereoisomers. In the event, these reactions deliver the corresponding cyclopropanes **3-3ag-3-3ai** with varying levels of diastereoselectivity depending on the substituent size, from vinyl (73%, 2.5:1 *d.r.*, major to Σ minor), to methyl (69%, 7.1:1 *d.r.*) and phenyl (62%, >20:1 *d.r.*). Using *trans*-1,2-disubstituted secondary allylic alcohols, we observed single diastereomers of cyclopropanes **3-3aj-3-3al** ranging in good to excellent yields.

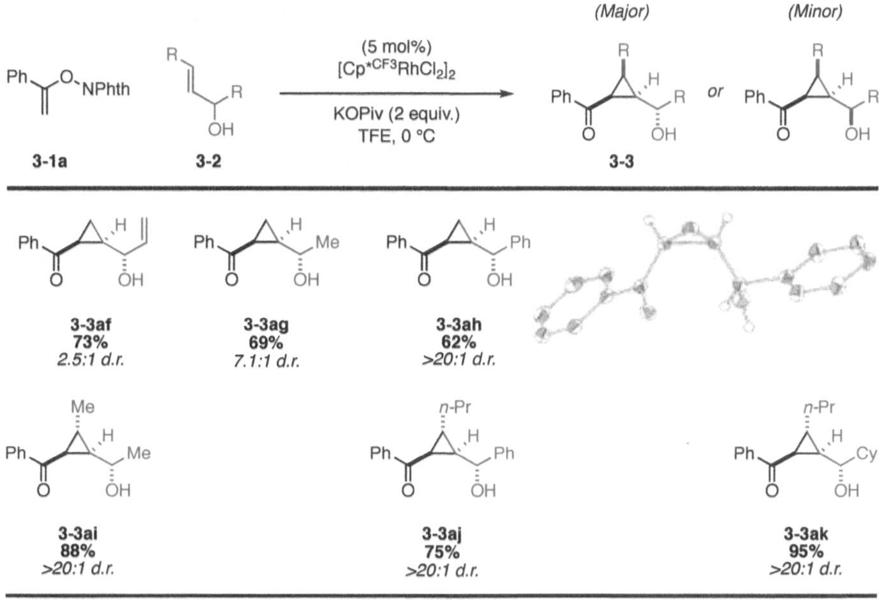

Scheme 3.4 Scope of secondary allylic alcohols.

Interestingly, when our cyclopropanation protocol was applied to secondary, cyclic allylic alcohols, the results revealed a divergence in the mechanism. Using hexenol with **3-1a** under the standard reaction conditions, the corresponding cyclopropane product was not observed. However, when cyclooctenol was used the corresponding cyclopropane was observed in 85% yield as a single diastereomer. This set of experiments revealed a few points about the mechanism of this reaction. On the basis of the crystal structure of **3-3ah**, we confirmed the *anti*-addition of the carbene transfer. Under Simmons-Smith reaction conditions, similar selectivities are observed for *syn*-addition to cyclohexenol and *anti*-addition to cyclooctenol. It is hypothesized using cyclooctenol, the eight-membered ring prefers to adapt a chair-boat confirmation with the complexed hydroxyl group in the equatorial position, allowing methylene transfer to easily be delivered to the closes face of the alkene in *anti* fashion as a single diastereomer. In our system, we propose the hydroxyl group is not directly complexed to the metal; however, since it is linked through phthalimide opening, the methyle transfer still prefers *anti*-addition.

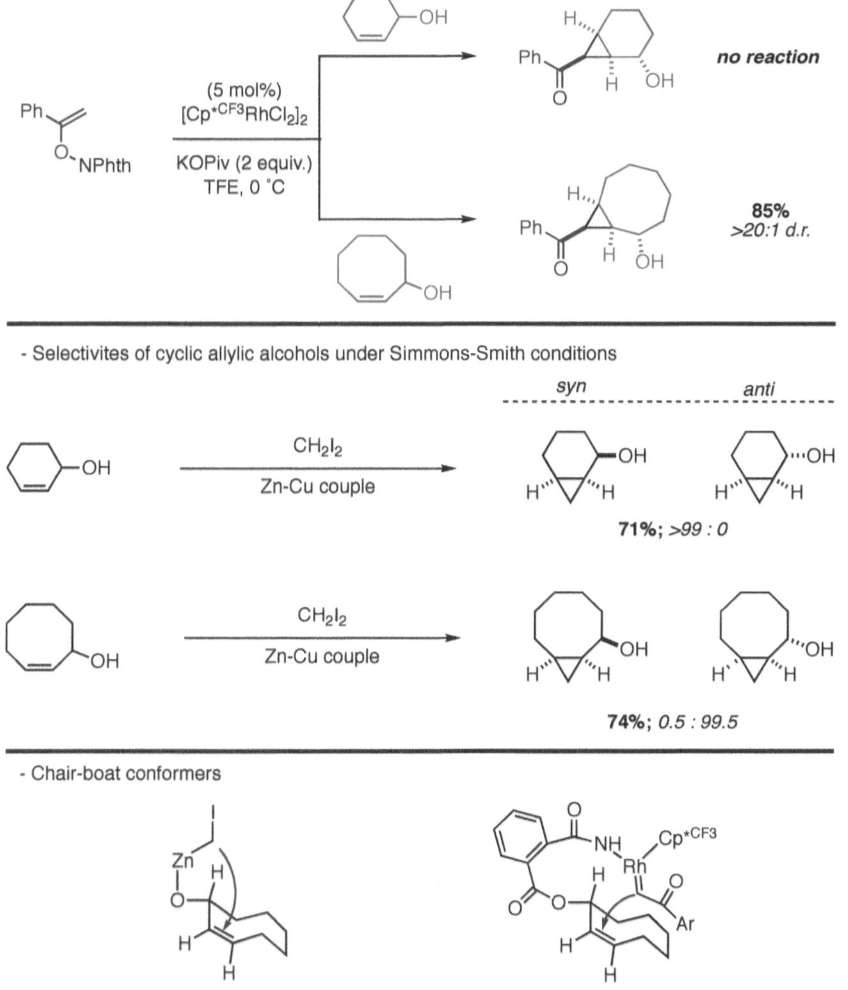

Figure 3.8 Comparison of secondary, cyclic allylic alcohols.

3.5 Mechanistic Studies

To interrogate the mechanism of this cyclopropanation reaction (Scheme 5), we first tested the length of the nucleophilic tether. Homoallylic alcohol **3-4a** gives cyclopropane **3-5aa** in only 12% yield indicating the chain length from the oxygen atom to the olefin is of great importance. Similarily, *bis*-homoallylic alcohol **3-6a** gives cyclopropane **3-7aa** in only 17% yield.

To showcase the regio-preference of our cyclopropanation protocol, **3-1a** was subjected to substrate **3-2m** (geraniol) bearing a tethered tri-substituted alkene as a potential competitive site for cyclopropanation. Gratifyingly, cyclopropane **3-3am** was generated in 55% yield with good diastereoselectivity and excellent regioselectivity. Nerol, the *cis* isomer, also gave the desired cyclopropane **3-3an** in lower yield but similar selectivities to geraniol. With these studies, we conclude that our cyclopropanation protocol is regioselective.

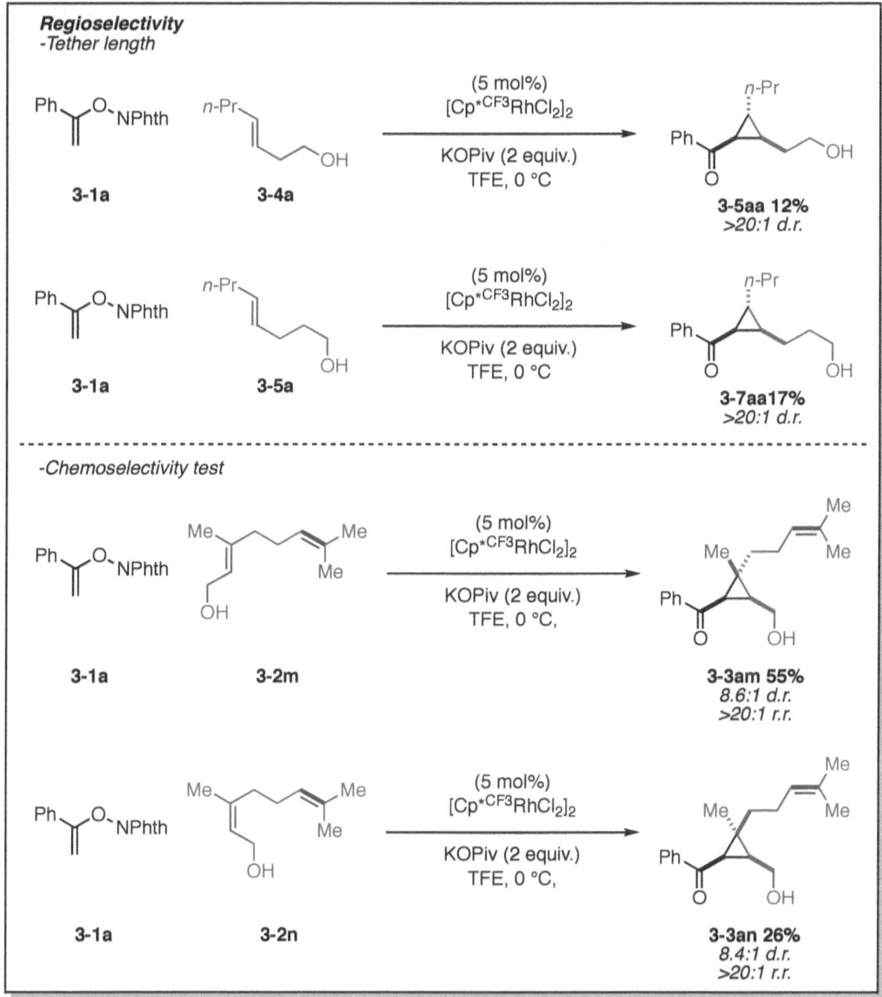

Figure 3.9 *Regioselective applications of the cyclopropanation protocol.*

Next, we sought to test if the tethered nucleophile was needed for reactivity. Allylic ether **3-8a** is a poor substrate with only trace **3-9aa** observed indicating the presence of an unhindered hydroxyl-group is necessary for the reaction to take place. Allylic carboxylic acid **3-8b** gives cyclopropane **3-9ab** in trace yield. Interestingly, protected allylic amine **3-8c** gives cyclopropane **3-9ac** in 77% yield and 9.5:1 *d.r.* From these experiements, we conclude that a nucleophilic attack is necessary for reactivity. The lack of reactivity with allylic ethers clearly shows this. Trace formation of **3-9ab** could be due to reduced nucleophilicity of carboxylates; however, addition of a carboxylic acid buffers the solution. From optimization reactions, 2 equivalents of base is needed for this transformation to proceed. When reactivity is restored using a pendant sulfonamide, we believe this functional group is nucleophilic enough to open the phthalimide ring and does not affect the concentration of base present in the reaction.

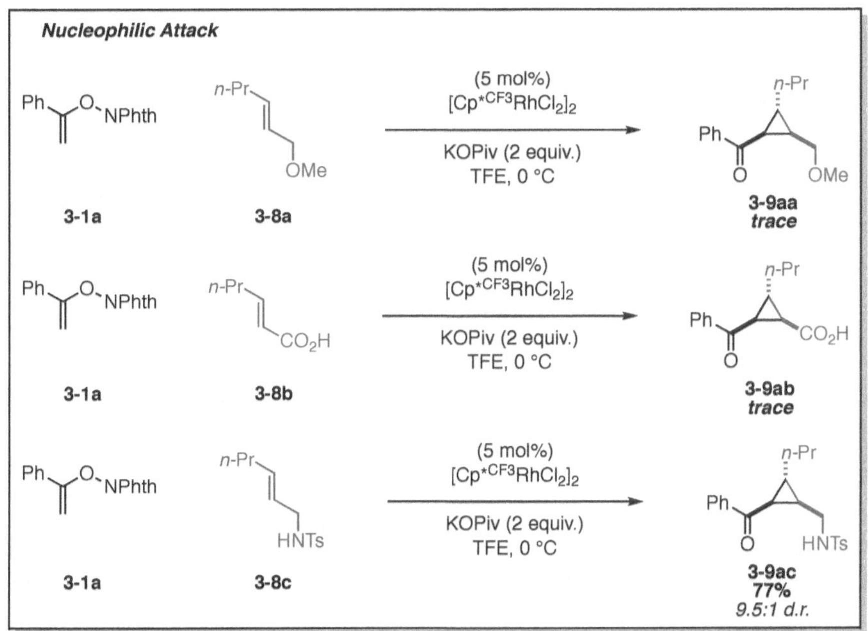

Figure 3.10 Investigations of the nucleophilicity of the allylic functional group.

We next subjected **3-1a** to the reaction conditions in the absence of alkene with TFE-d_1 solvent and observed no deuteration of the alkenyl protons suggesting that the C–H activation is irreversible. Using a deuterium labeled allylic alcohol at the alkene, we again observe a stereospecific transformation as the desired cyclopropane is observed in 82% yield while the proton and deuteron maintain complete *trans* relationship from the starting alkene.

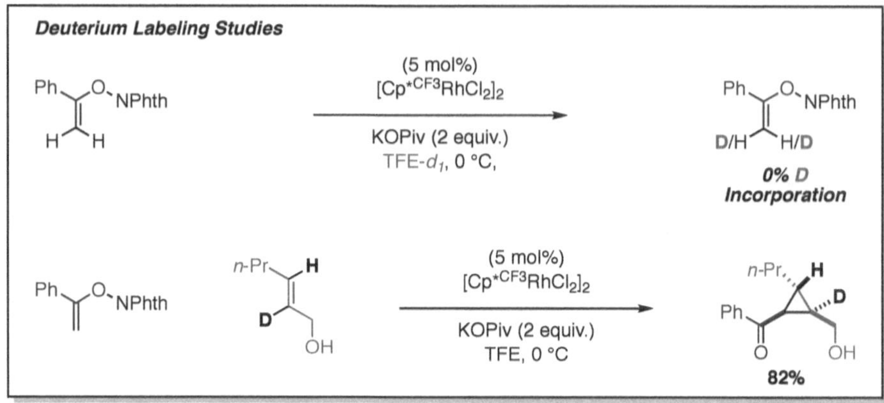

Figure 3.11 Deuterium labeling studies.

In another experiment, we set out to detect potential reactivity between **3-1a** and **3-2b** in the absence of Rh catalyst and we were surprised to observe the formation of dioxazoline **3-10ac** in 38% yield with 1 equivalent of KOPiv in THF at room temperature. We speculate this occurs via opening of the phthalimide ring and acylation of the allylic alcohol (eq. 8). Subjecting dioxazoline **3-10ac** to the cyclopropanation reaction conditions did not afford cyclopropane, suggesting that dioxazoline **3-10ac** is an off-cycle product.

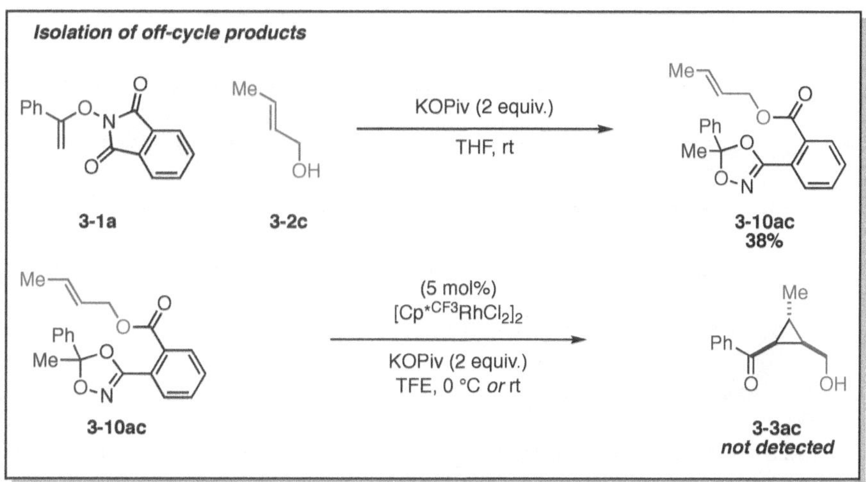

Figure 3.12 *Observation and intermediacy test of dioxazoline.*

Furthermore, dioxazoline **3-10ac** is observed while monitoring the reaction by crude ^{1}H-NMR (Appendix Two), indicating, that the phthalimide ring is opened during the reaction and not upon workup.

From these studies, we conclude that the cyclopropanation reaction: 1) is regioselective, 2) is conformationally dependent, 3) requires a tethered nucleophile, and 4) is initiated by an irreversible C–H activation.

3.6 Proposed Mechanism

On the basis of these experiments, we propose the following mechanism (Scheme 6). First, **3-2** undergoes acylation with **3-1** that gives intermediate **I**. Maintaining the reaction temperature at 0 °C inhibits cyclization to afford the dioxazoline product, which is instead intercepted by the active Rh(III) catalyst **II**. Intermediate **I** undergoes N–H deprotonation that gives intermediate **III** to initiate an irreversible C–H activation via concerted metalation-deprotonation that results in rhodacycle **IV**. At this stage, we hypothesize the formation of intermediate **V** by cleavage of the N–O bond and formation of a Rh-carbene. Due to the prior acylation of the allylic alcohol, intermediate **VI** is formed via the [2+1] annulation where the Rh-carbene is delivered across the alkene and on the same face as the pendent oxygen atom in stereoselective fashion. Protodemetallation and subsequent phthalimide ring closure releases the product and turns the catalyst over.

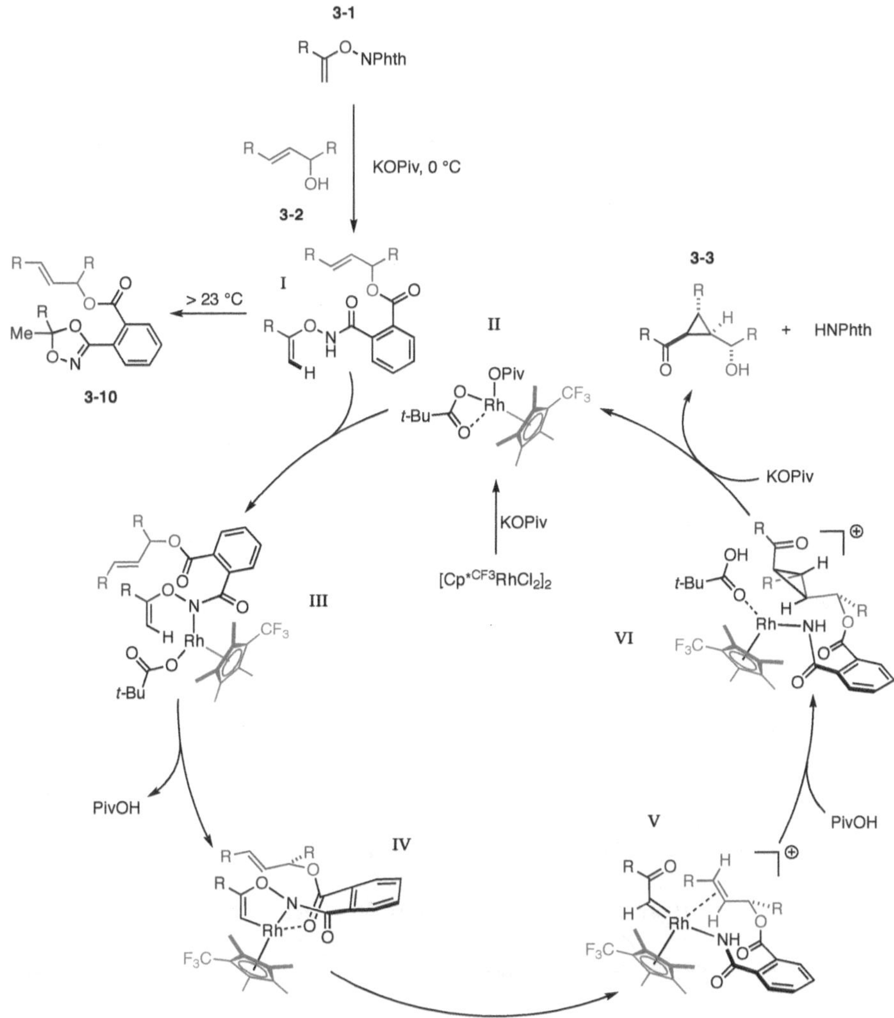

Figure 3.13 *Proposed mechanism.*

3.7 Summary

In conclusion, we have developed a directed diastereoselective cyclopropanation protocol for the [2+1] annulation of *N*-enoxyphthalimides and allylic alcohols. The diastereoselectivity of the reaction is speculated to arise from an intermediate generated by a ring-opening acylation of the allylic alcohol. Generation of a Rh-carbenoid leads to intramolecular cyclopropanation in excellent yield and diastereoselectivity.

3.8 References

(1) (a) Chen, D.Y.-K.; Pouwer, R.H.; Richard, J.-A. *Chem. Soc. Rev.* **2012**, *41*, 4631. (b) Talele, T. T. *J. Med. Chem.* **2016**, *59*, 8712.

(2) (a) Doyle, M. P.; Forbes, D. C. *Chem. Rev.* **1998**, *98*, 911. (b) Lebel, H.; Marcoux, J.-F.; Molinaro, C.; Charette, A. B. *Chem. Rev.* **2003**, *103*, 977.

(3) For selected recent references, see: (a) Lindsay, V. N. G.; Lin, W.; Charette, A. B. *J. Am. Chem. Soc.*, **2009**, *131*, 16383. (b) Lindsay, V. N. G.; Nicolas, C.; Charette, A. B. *J. Am. Chem. Soc.* **2011**, *133*, 8972. (c) Negretti, S.; Cohen, C. M.; Chang, J. J.; Guptill, G. M.; Davies, H. M. L. *Tetrahedron* **2015**, *71*, 7415. (d) Lehner, V.; Davies, H. M. L.; Reiser, O. *Org. Lett.* **2017**, *19*, 4722. (e) Tindall, D. J.; Werlé, C.; Goddard, R.; Philipps, P.; Farès, C.; Fürstner, A. *J. Am. Chem. Soc.* **2018**, *140*, 1884.

(4) For selected recent references, see: (a) Chanthamath, S.; Iwasa, S. *Acc. Chem. Res.* **2016**, *49*, 2080. (b) Maas, G. *Chem. Soc. Rev.* **2004**, *33*, 183.

(5) For selected recent references, see: (a) Taber, D. F.; Amedio, J. C.; Sherrill, R. G. *J. Org. Chem.* **1986**, *51*, 3382. (b) Denmark, S. E.; Stavenger, R. A.; Faucher, A.-M.; Edwards, J. P. *J. Org. Chem.* **1997**, *62*, 3375. (c) Chen, S.; Ma, J.; Wang, J. *Tetrahedron Lett.* **2008**, *49*, 6781.

(6) (a) Nozaki, H.; Takaya, H.; Moriuit, S.; Noyori, R. *Tetrahedron* **1968**, *24*, 3655. (b) Salomon, R. G.; Kochi, J. K. *J. Am. Chem. Soc.* **1973**, *95*, 3300.

(7) (a) Huang, L.; Chen, Y.; Gao, G.-Y.; Zhang, X. P. *J. Org. Chem.* **2003**, *68*, 8179. (b) Chen, Y.; Fields, K. B.; Zhang, X. P. *J. Am. Chem. Soc.* **2004**, *126*, 14718. (c) Chen, Y.; Zhang, X. P. *J. Org. Chem.* **2007**, *72*, 5931. (d) Chen, Y.; Ruppel, J. V.; Zhang, X. P. *J. Am. Chem. Soc.* **2007**, *129*, 12074. (e) Zhu, S.; Ruppel, J. V.; Lu, H.; Wojtas, L.; Zhang, X. P. *J. Am. Chem. Soc.* **2008**, *130*, 5042.

(8) (a) Hamaker, C. G.; Mirafzal, G. A.; Woo, L. K. *Organometallics* **2001**, *20*, 5171. (b) Aggarwal, V. K.; de Vicente, J.; Bonnert, R. V. *Org. Lett.* **2001**, *3*, 2785. (c) Coelho, P. S.; Brustad, E. M.; Kannan, A.; Arnold, F. H. *Science* **2013**, *339*, 307. (d) Allouche, E. M. D.; Al-Saleh, A.; Charette, A. B. *Chem. Commun.* **2018**, *54*, 13256.

(9) Ebner, C.; Carreira, E. M. *Chem. Rev.* **2017**, *117*, 11161.

(10) (a) Dehmlow, E. V.; Stütten, J. *Tetrahedron Lett.* **1991**, *32*, 6105. (b) Charette, A. B.; Molinaro, C.; Brochu, C. *J. Am. Chem. Soc.* **2001**, *123*, 12168. (c) Bull, J. A.; Charette,

A. B. *J. Am. Chem. Soc.* **2010**, *132*, 1895. (d) Allouche, E. M. D.; Taillemaud, S.; Charette, A. B. *Chem. Commun.* **2017**, *53*, 9606. (e) Benoit, G.; Charette, A. B. *J. Am. Chem. Soc.* **2017**, *139*, 1364. (f) Werth, J.; Uyeda, C. *Angew. Chem. Int. Ed.* **2018**, *57*, 13092.

(11) Piou, T.; Rovis, T. *J. Am. Chem. Soc.* **2014**, *136*, 11292.

(12) Piou, T.; Romanov-Michailidis, F.; Ashley, M. A.; Romanova-Michaelides, M.; Rovis, T. *J. Am. Chem. Soc.* **2018**, *140*, 9587.

(13) This reaction has recently been rendered asymmetric by Cramer and coworkers; see: Duchemin, C.; Cramer, N. *Chem. Sci.* **2019**, *10*, 2773.

(14) Piou, T.; Rovis, T. *Nature* **2015**, *527*, 86.

(15) Duchemin, C.; Cramer, N. *Org. Chem. Front.* **2019**, *6*, 209.

– Chapter Four –

Validating Isolated Reaction Intermediates
for 1,1-Carboamination of N-enoxyphthalimides

4.1 Artifacts of the Cyclopropanation Reaction

While probing the scope of cyclopropanation of unactivated alkenes, we subjected **4-1a** to the standard reaction conditions under an atmosphere of ethylene. We found complete consumption of starting material but only trace desired cyclopropane **4-3aa**.

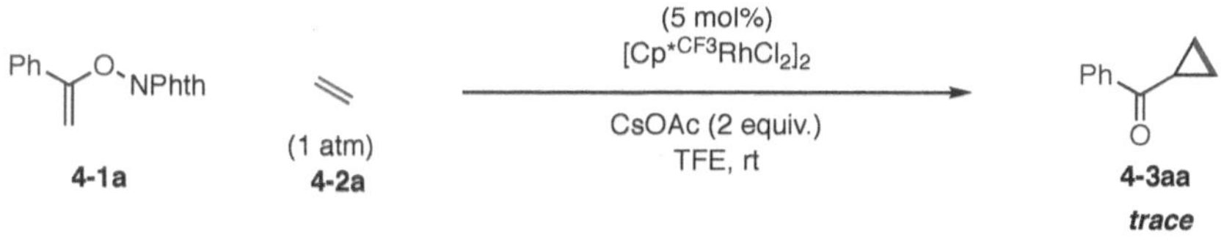

Figure 4.1 *Cyclopropanation reaction with ethylene as the alkene.*

After purification, we observed the formation of a Rh-π-allyl complex in 83% yield. This was further confirmed by X-ray crystallography.

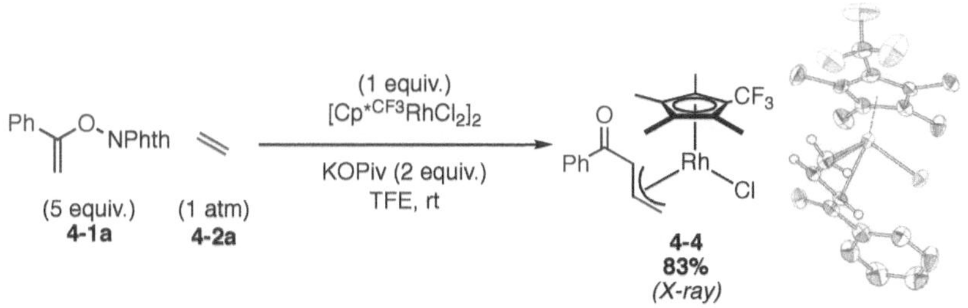

Figure 4.2 *Formation of Rh-π-allyl complex.*

Similar to the mechanism described in chapters 2 and 3, we believe **4-4** is provided

by CMD-type C–H activation of *N*-enoxyphthalimides. After migratory insertion, a 7-

membered rhodacycle is likely formed. From here, the π-allyl species observed can be

furnished by a number of transformations. The pathway we favor involves a beta-hydride

elimination that gives a Rh-hydride that can undergo sigma bond metathesis to cleave

the N–O bond. After Rh-enolate isomerization and ligand substitutions, **4-4** is formed.

Overall, this is a redox-neutral process, so the Rh-center may remain Rh(III) at all times.

However, the Rh could undergo earlier oxidation (via cleavage of the N–O bond) or

reduction (likely by deprotonation of a metal-hydride).

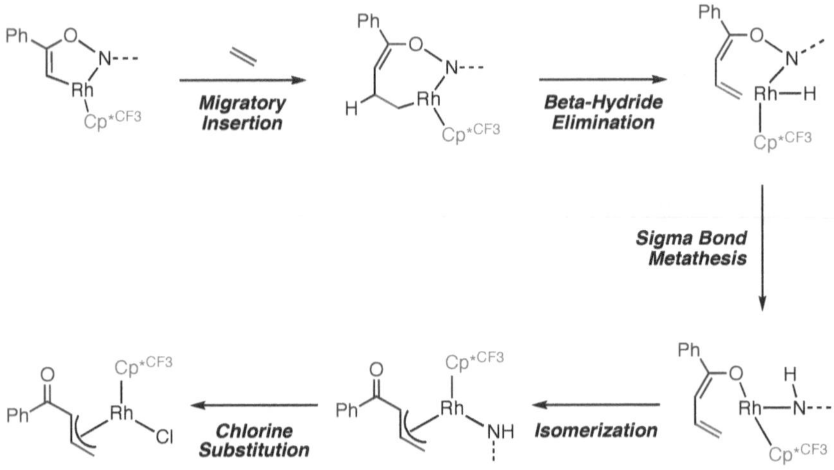

*Figure 4.3 Likely pathway for the formation of **4-4**.*

We first wanted to evaluate if **4-4** played a role in the cyclopropanation of

unactivated alkenes. To test this, we subjected **4-1a** and **4-2b** to 5 mol% of **4-4** in the presence of CsOAc and TFE at room temperature. We found that cyclopropane **4-3ab** is afforded in 9% yield, which is dramatically lower than the μ-dichloride precatalyst.

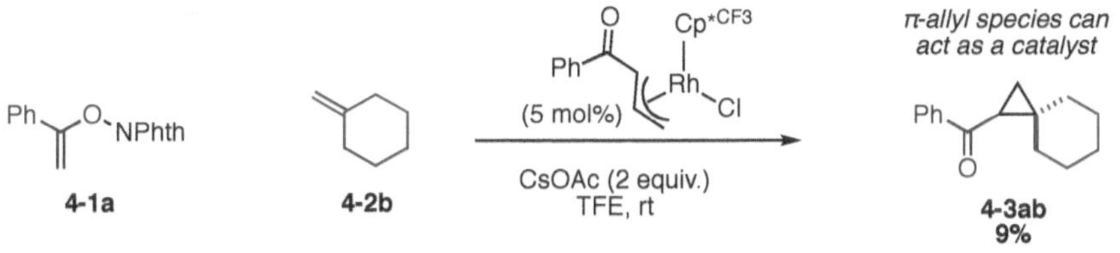

Figure 4.4 Subjection of 4-4 to cyclopropanation reaction conditions.

While we observed turnover with **4-4** as the catalyst in the cyclopropanation of **4-2a**, we wanted to leverage the formation of a π-allyl species to furnish a new bond.

4.2 Overview of C–N Bond Formation from π-Allyl Species from Nitrenoid Precursors

Building on previous success in $Cp^XIr(III)$-catalyzed transformations,[1] our group first reported the branched-selective allylic amination of terminal alkenes from dioxazolones. In the presence of LiOAc, Ag-salt, and Cp*Ir(III) catalyst, terminal alkenes undergo allylic C–H activation to afford η-3 Ir-π-allyl complexes. Importantly, the isolable π-allyl complexes can be converted to the desired allylic amide when subjected to dioxazolone.[2] It is suggested that the Ir-π-allyl complex is oxidized by cleavage of the N–O bond that affords a Ir-nitrene. The resulting Ir(V)-nitrene then undergoes fast reductive elimination and protodemetallation.

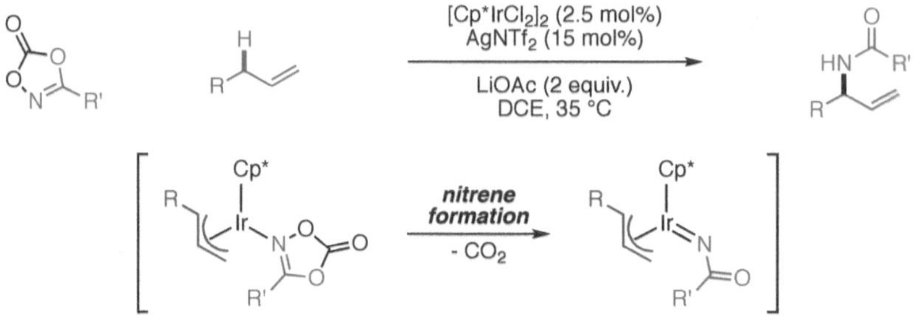

Figure 4.5 Ir(III)-catalyzed intermolecular branched-selective allylic amination of terminal alkenes.

Soon after, Glorius and coworkers expanded the scope to include internal alkenes with minimal changes to the reaction conditions.[3]

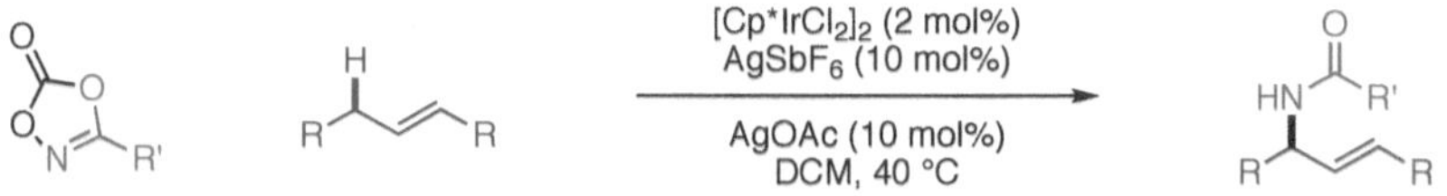

Figure 4.6 Intermolecular amination of internal alkenes.

Blakey and coworkers also reported on the Ir-catalyzed branched-selective allylic amination in a report the same year.[4] Interestingly, they found that in the presence of catalytic CsOAc and Ag-salt in DCE at 40 °C, simply switching from an Ir to a Rh precatalyst diverted the outcome to afford linear-selective allylic amination of alkenes.

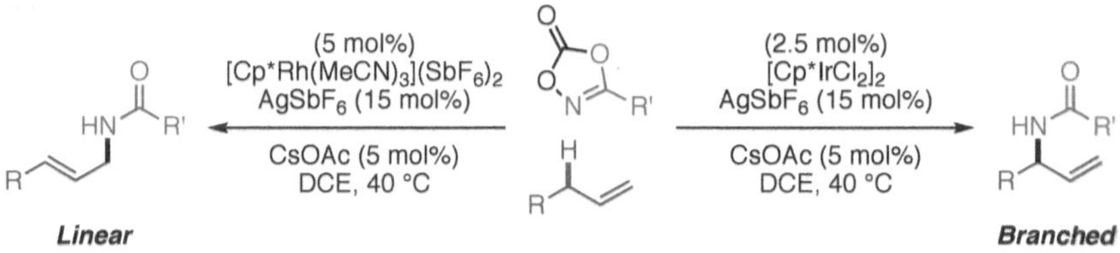

Figure 4.7 Catalyst-dependent regioselective allylic amination of alkenes.

We looked to these previous successes of our group's and others to provide new reactivity for the Rh-π-allyl complex **4-4**.

4.3 Envisioned 3-component Reaction

Naturally, the idea we gravitated towards was subjecting a dioxazolone to the conditions established for the synthesis of **4-4**. We would expect to see a number of regioisomers, with **4-6**, the branched-selective terminal alkene, dominating (drawn below).

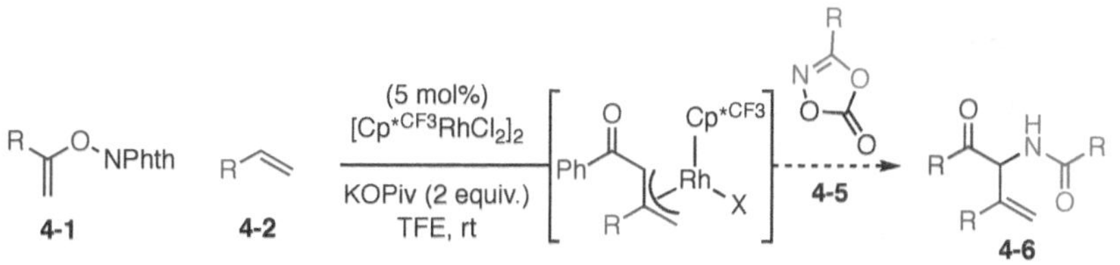

Figure 4.8 Proposed Rh(III)-catalyzed 3-component 1,1-carboamination of N-enoxyphthalimides.

From a mechanistic standpoint, what we envisioned was a merger between the cyclopropanation chemistry and the allylic amination chemistry. We believe the formation of intermediate III is the same as previously described. At this point, insertion of **4-2** would afford Rhodacycle IV and most likely undergo beta-hydride elimination that gives a Rh-hydride diene complex. Intermediate V then affords π-allyl complex VI with dioxazolone bound. After oxidation, Rh-nitrene, intermediate VII, undergoes facile reductive elimination to afford **4-6** after protodemetallation turns over the catalyst.

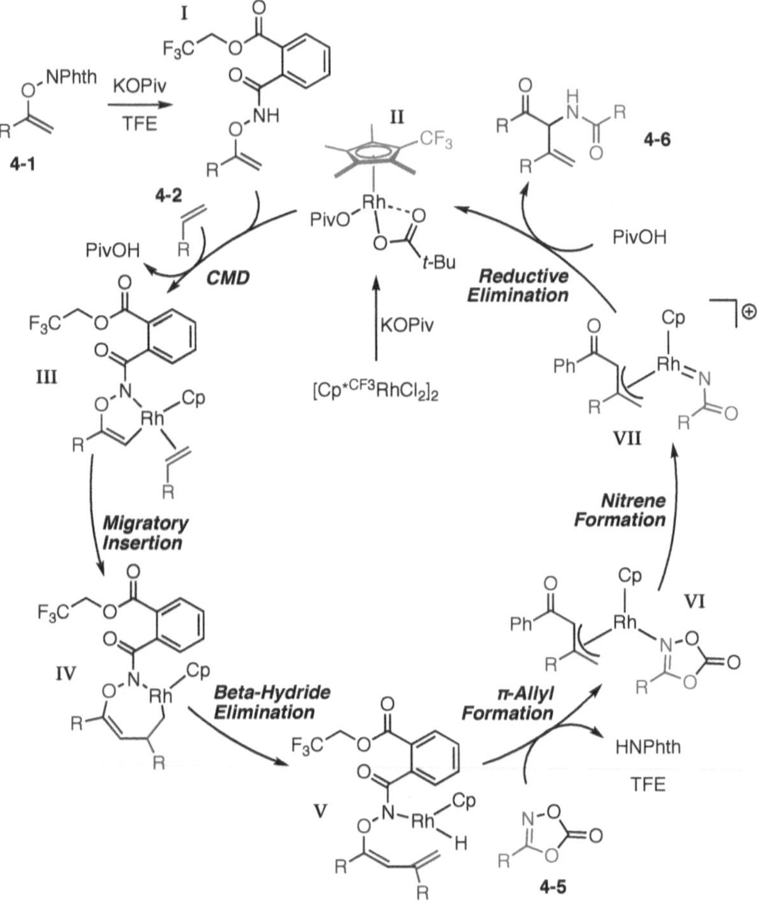

Figure 4.9 *Envisioned mechanism of 1,1-carboamination of N-enoxyphthalimides.*

4.4 Attempts at 3-component 1,1-Carboamination

Initially, we subjected **4-1a** under ethylene atmosphere to **4-5a** in the presence of Ag salt and Rh(III) catalysts. While none of the entries discussed provide the desired carboamination product, the consumption of **4-1a** was measured in each case in hopes of pushing the reactivity forward. With Cp* and Cp*CF3 and catalytic amounts of base, **4-1a** remains untouched. However, when stoichiometric amount of base are used as in Figure 4.2, degradation of **4-1a** is observed. Furthermore, removing the Ag additive causes almost complete consumption of **4-1a** with either precatalyst.

Entry	[Rh] precatalyst	Base Equiv.	Additive	4-1a Remaining
A	[Cp*RhCl$_2$]$_2$	0.2	AgSbF$_6$	100%
B	[Cp*CF3RhCl$_2$]$_2$	0.2	AgSbF$_6$	100%
C	[Cp*RhCl$_2$]$_2$	2	AgSbF$_6$	26%
D	[Cp*CF3RhCl$_2$]$_2$	2	AgSbF$_6$	56%
E	[Cp*RhCl$_2$]$_2$	2	none	10%
F	[Cp*CF3RhCl$_2$]$_2$	2	none	12%

Scheme 4.1 Initial reaction screening toward 1,1-carboaminaiton.

Next, we briefly screened other nitrenoid precursors. However, seeing no amination products we chose to push forward using dioxazolones as we thought these could be an easier oxidation than others we tried.

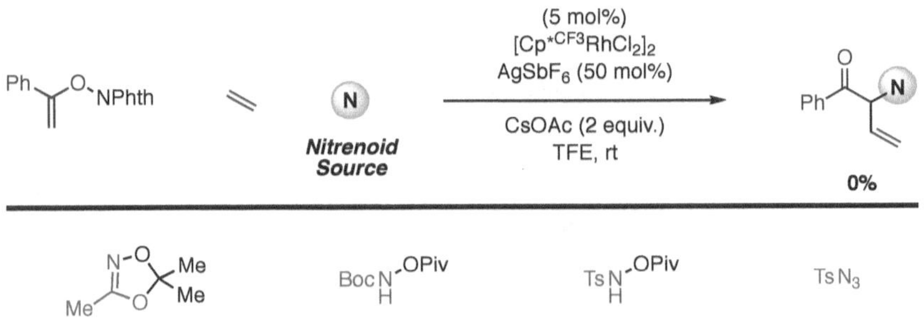

Scheme 4.2 *Screen of nitrenoid precursors.*

Facing a collection of results that showed catalysis was certainly not happening, we turned to stoichiometric studies to provide some answers.

4.5 Stoichiometric Studies

Typically, when prospecting for new reactivity in the realm of Rh(III)-catalyzed C–H functionalization, we begin using the parent Cp* ligand. While consumption of **4-1a** is observed under both Rh(III) precatalysts, π-allyl complex formation is not observed with [Cp*RhCl$_2$]$_2$ as the precatalyst. This result indicates that Cp* is not the correct ligand to initiate catalysis for this transformation.

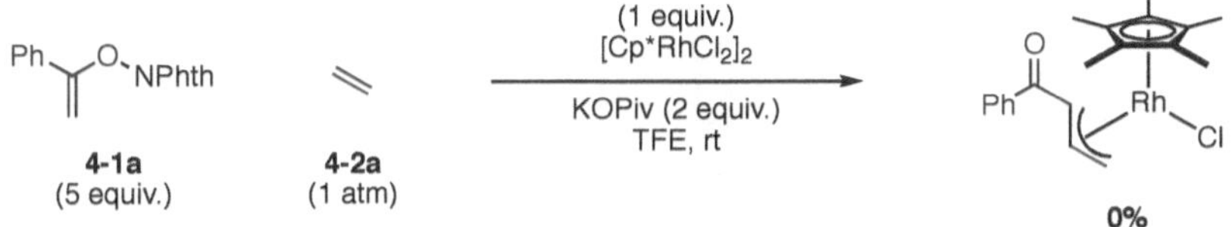

Figure 4.10 Attempted π-allyl complex synthesis with Cp* as a ligand.

Additionally, **4-4** was subjected to **4-5a** in the presence of KOPiv in TFE at room temperature where neither branched nor linear products are observed. This tells us that the reaction conditions optimal for the synthesis of **4-4** are not conducive to allylic C–N bond formation events. Furthermore, subjecting **4-4** to **4-5a** in the presence of 2 equivalents of a Ag salt in DCE at 40 °C–reaction conditions similar to the allylic amination in Figure 4.7–similarly results in no desired product observation. This further demonstrates that catalyst selection is even more important. While Cp*[CF3] enables formation of **4-4**, this ligand is not suitable for the C–N bond forming event.

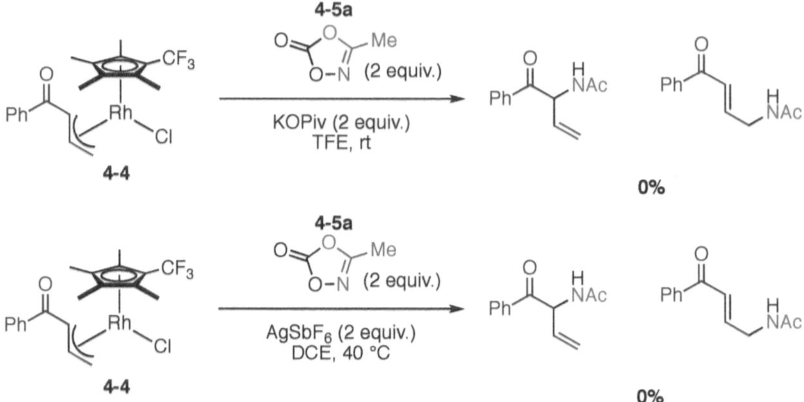

Figure 4.11 *Attempts at C–N bond formation from π-allyl precursors.*

From these stoichiometric studies, clearly π-allyl complex **4-4** does not provide the desired amination reaction as other unactivated alkenes provide. As of now, we believe it due predominately to electronics of the catalyst. From a follow-up report on allylic amination of nearly identical C–H bonds, the amination occurs at the more electron-rich carbon. The electron-deficient substituent resulting from employing **4-1** as starting material imparts a large electronic effect on the substrate. Furthermore, the Cp ligand probably needs to be electron deficient to afford the Rh-π-allyl complex. However, we believe that for Rh(III) species to undergo 2e⁻ oxidation, the metal center would prefer to be as electron-rich as possible. This step should favor electron-rich Cp ligands, as demonstrated in Ir chemistry. This means that these potential 2 key steps lay at odds with one-another

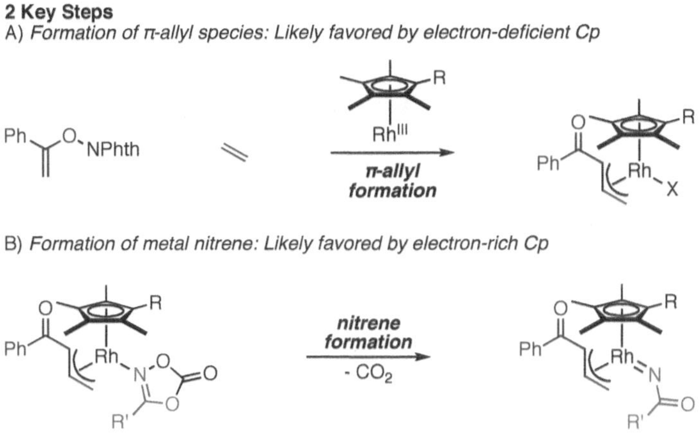

Figure 4.12 *Potential catalyst incompatibility of key steps involved in 1,1-carboamination.*

4.6 Summary

We have demonstrated that *N*-enoxyphthalimides, in the presence of [Cp*^{CF3}RhCl$_2$]$_2$ complex under ethylene atmosphere, undergo efficient synthesis of a Rh-π-allyl complex. Based on previous success centered around the design of π-allyl intermediates, we envisioned the construction of C–N bond would occur when subjected to a nitrenoid precursor. Attempts to catalyze a 1,1-carboamination reaction were made with a variety of conditions in the presence of different Rh precatalysts. From stoichiometric studies, we can conclude that the choice of catalyst is of utmost priority to unlock new reactivity.

4.7 References

____

(1)   a) Lei, H.; Conway, J. H.; Cook, C. C.; Rovis, T. *J. Am. Chem. Soc.* **2019**, *141*, 11864.

b) Conway, J. H.; Rovis, T. *J. Am. Chem. Soc.* **2018**, *140*, 135. c) Romanov-Michailidis, F.; Ravetz, B. D.; Paley, D. W.; Rovis, T. *J. Am. Chem. Soc.* **2018**, *140*, 5370. d) Lei, H.; Rovis, T. *Nat. Chem.* **2020**, *12*, 725.

(2)   Lei, H.; Rovis, T. *J. Am. Chem. Soc.* **2019**, *141*, 2268.

(3)   Knecht, T.; Mondal, S.; Ye, J.-H; Das, M.; Glorius, F. *Angew. Chem. Int. Ed.* ***2019**, *58*, 7117.

(4)   Burman, J. S.; Harris, R. J.; Farr, C. M. B.; Bacsa, J.; Blakey, S. *ACS Catalysis* **2019**, *9*, 5474

– Chapter Five –

Rh(III)-catalyzed 1,2-Carboamination of Alkenes via sp³ C–H Activation

5.1 Introduction to 1,2-Carboamination

In 2015, our group reported that under Rh(III) catalysis, 1,2-disubtituted alkenes undergo *syn*-1,2-carboamination using *N*-enoxyphthalimides.[1] Notably, the phthalimide handle is incorporated in the product acting as a traceless directing group for C–H activation as well as an internal oxidant, making this an incredibly efficient process. This is achieved by modifying the Cp ligand on Rh and using methanol as solvent, where *N*-enoxyphthalimides–previously known to facilitate cyclopropanation chemistry–experience a chemoselective transformation.

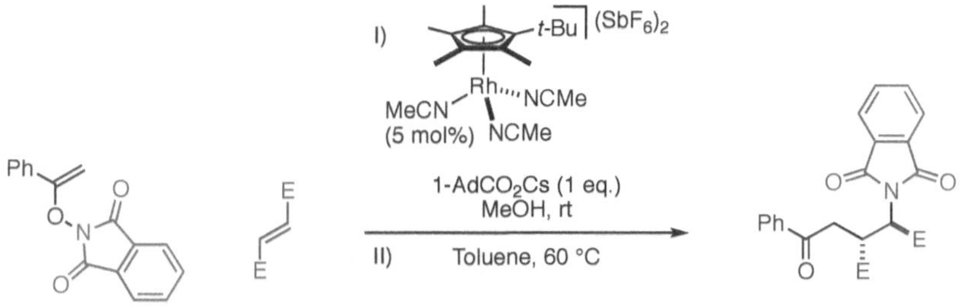

Figure 5.1 Rh(III)-catalyzed syn-1,2-carboamination of fumarate-type alkenes.

Mechanistically, methanol is proposed to open the phthalimide ring revealing a bidentate directing group. After C–H activation, fumarate type alkenes undergo migratory insertion that give a 7-membered rhodacycle. From here, C–N bond formation is proposed via a reductive pathway that yields a Rh(I) complex that is turned over by oxidative addition of the N–O bond. Alternatively, C–N bond formation could occur through an oxidative pathway via nitrene formation followed by reductive elimination. Removal of methanol and heating in toluene results in phthalimide ring closure to furnish the desired carboamination product.

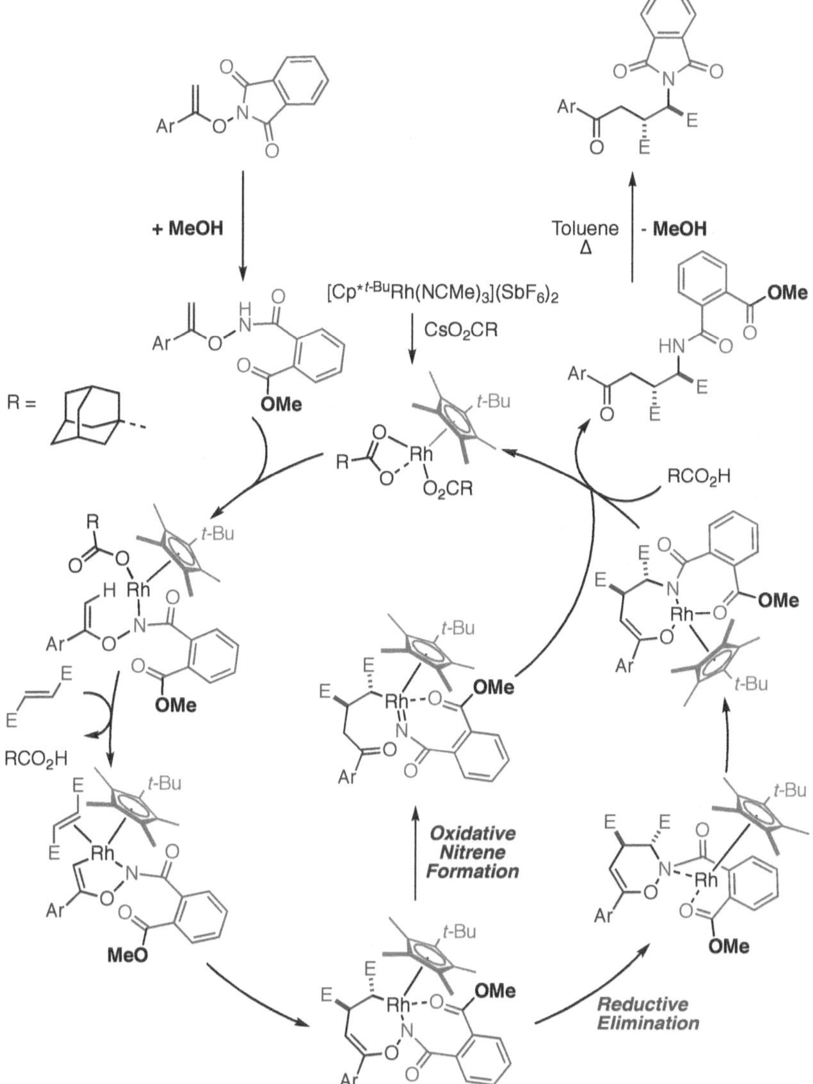

Figure 5.2 *Proposed mechanism of 1,2-carboamination of alkenes from N-enoxyphthalimides.*

Notably, this reaction has recently been rendered enatioselective by Cramer and coworker using *N*-enoxysuccinimides.[2] 1,2-carboamination of electron-rich alkenes using *N*-enoxyphthalimides has also been reported under photoredox catalysis.[3]

5.2 Substrates Beyond *N*-enoxyphthalimides

While *N*-enoxyphthalimides constitute valuable starting materials for the construction of C–C and C–N bonds, they come with a large downside of heavy pre-functionalization. Anderson and coworkers first reported their synthesis over 4 steps from styrenes.[4] Recently, Cramer and coworker disclosed a concise 1-step alternative to the synthesis of *N*-enoxyimides from terminal alkynes under Au catalysis.[5] Importantly, this pathway included the ability to produce alkyl substituted *N*-enoxyimides.

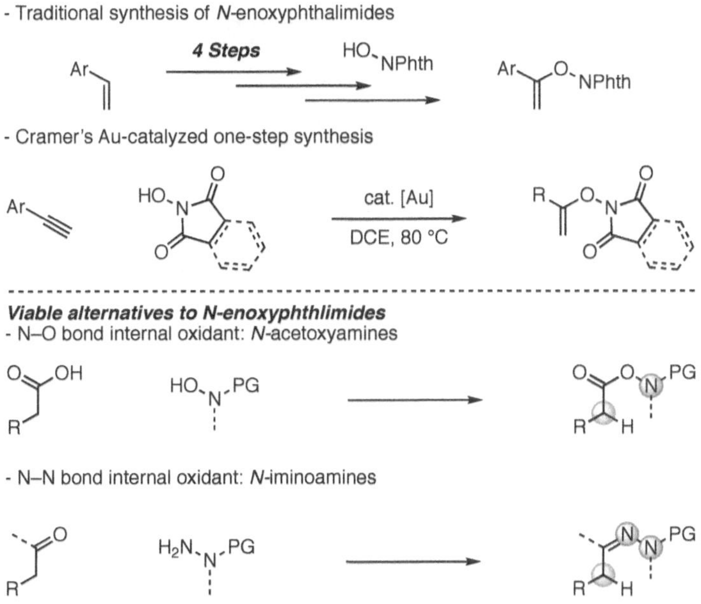

Figure 5.3 *N-enoxyphthalimide synthesis and potential alternatives.*

As we began to ponder alternatives to N-enoxyphthalimides, three factors need to be considered. Alternatives must have: 1) a nitrogen directing group, 2) a nitrogen-heteroatom bond as an internal oxidant, and 3) simple starting materials and easy synthetic routes. Two candidates emerged as viable starting points: N-acetoxyamines, that come from esterification of carboxylic acids, and N-iminoamines, that are the product of condensation between ketone/aldehydes and protected hydrazines.

5.3 Envisioned Mechanism from *N*-acetoxyamines

Beginning our studies around *N*-acetoxyamine **5-1**, we believe under Rh(III)-catalysis the nitrogen could direct the Rh to activate the sp^3 C–H bond alpha to the carbonyl, also taking advantage of the lowered pKa of the bond. If C–H bond activation occurs, rhodacycle II may insert an alkene **5-2** that gives 7-membered rhodacycle III. As proposed previously, the C–N bond may be furnished via an oxidative or reductive mechanism, where both pathways meet at intermediate V. Protodemetallation would then turn over the catalyst and give desired carboxylic acid **5-3**.

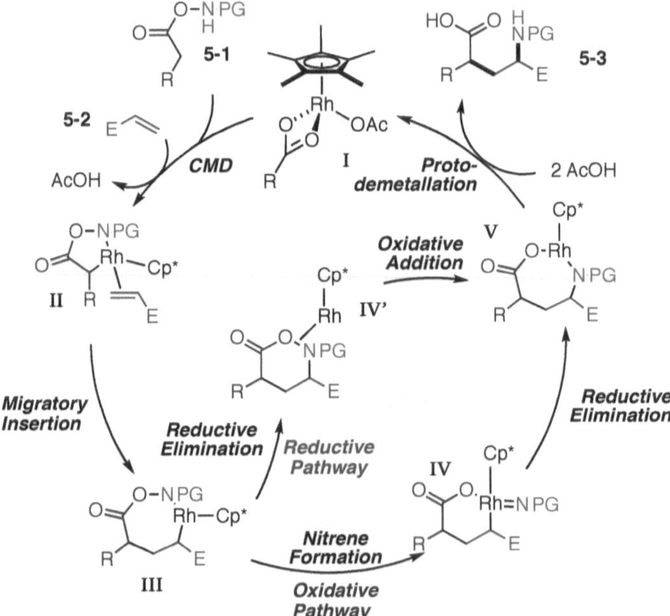

Figure 5.4 Predicted pathways for 1,2-carboamination of alkenes from N-acetoxyamines.

5.4 Carboamination of Alkenes from *N*-acetoxyphthalimides

We began our studies using *N*-acetoxyphthalimide **5-1a** with alkene **5-2a** in the presence of Cp*Rh(III)-precatalyst, KOAc in methanol at varying temperatures. We were pleased to find MeOH is capable of phthalimide ring opening in each case. At the time, we believed we were seeing the formation of **5-3aa** in 14% yield (entries A-C). Buffering the system with 1 equiv. of acetic acid caused the yield to decrease to 7% (entry D). Varying base between catalytic (0.2 equiv.) and super-stoichiometric (2.5 equiv.) also show slight depression in yield of **5-3aa** (entries E and F).

Entry	Base equiv.	Temp	Additive	**5-3aa** yield
A	1	rt	none	trace
B	1	40 °C	none	14%
C	1	65 °C	none	7%
D	1	40 °C	AcOH	8%
E	0.2	40 °C	none	10%
F	2.5	40 °C	none	7%

Scheme 5.1 Carboamination screens in methanol.

After a brief solvent screen, we were surprised to see that TFE was also leading to small amounts of **5-3aa** (entry G) while other solvents were incapable of activating **5-1a** for functionalization (entries H-J). Similar to the cyclopropanation reaction conditions, super-stoichiometric amounts of KOAc showed an increase in yield to 18% (entry K). Furthermore, Ag salt additive to render cationic Rh-species saw **5-3aa** rise to 24% yield.

Entry	Base equiv.	Solvent	Additive	*5-3aa* yield
G	1	TFE	none	7%
H	1	DCE	none	0%
I	1	THF	none	0%
J	1	HFIP	none	0%
K	2.5	TFE	none	18%
L	2.5	TFE	AgSbF$_6$	24%

Scheme 5.2 Solvent screen leading to TFE conditions.

At this moment in optimizations, we judged the formation of **5-3aa** on a doublet of doublet of doublet signal in the ^{1}H-NMR from δ 4.43-4.46. Unfortunately, upon attempted isolation of the acid, we found the formation of **5-4aa** in its place. While this is a carboamination, it is selective for the *sp²* C–H functionalization over the desired *sp³* C–H bond functionalization.

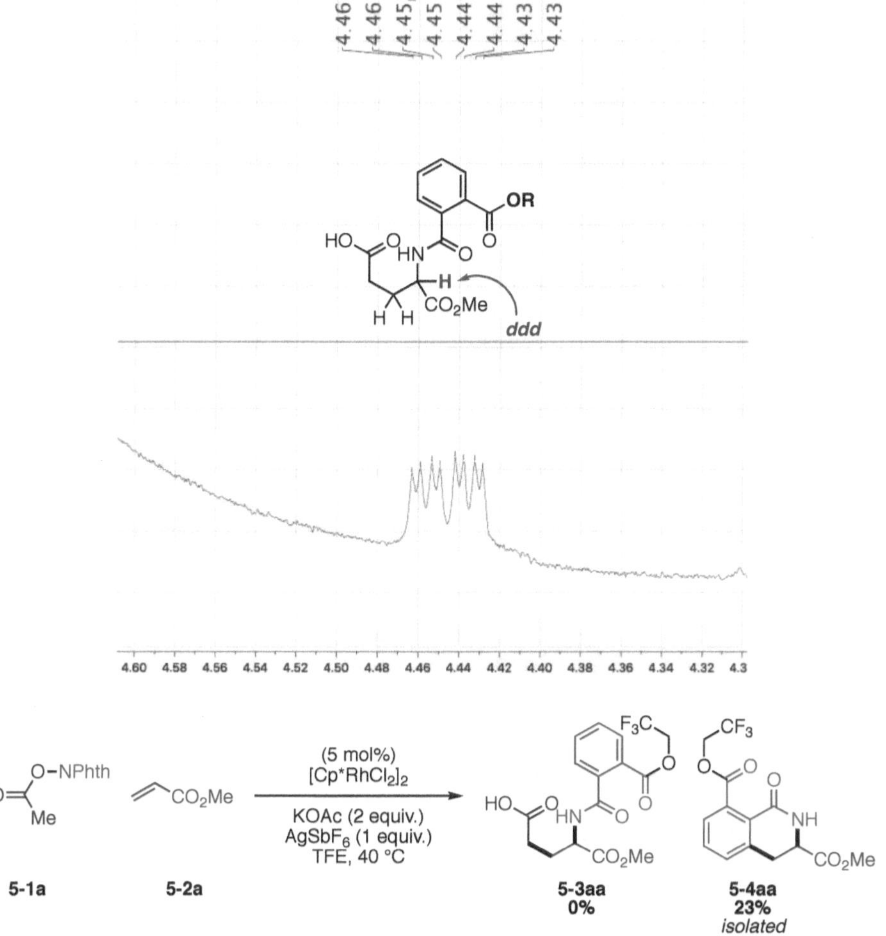

Figure 5.5 *Diagnostic ¹H-NMR signal and isolation of undesired byproduct.*

5.5 Future Considerations Concerning sp^3 C–H Functionalization of *N*-acetoxyphthalimides

We believe the phthalimide ring is being opened by alcoholic solvent and coordinating the Rh-catalyst with acetate ligand bound. From here, the catalyst discriminates between activating sp^2 or sp^3 C–H bonds. We believe that activation of the sp^2 C–h bond, while possessing a higher pKa, supplies a more stable rhodacycle. Concerning future studies, selection of a Cp ligand may be crucial in order to provide the desired sp^3 C–H functionalized products.

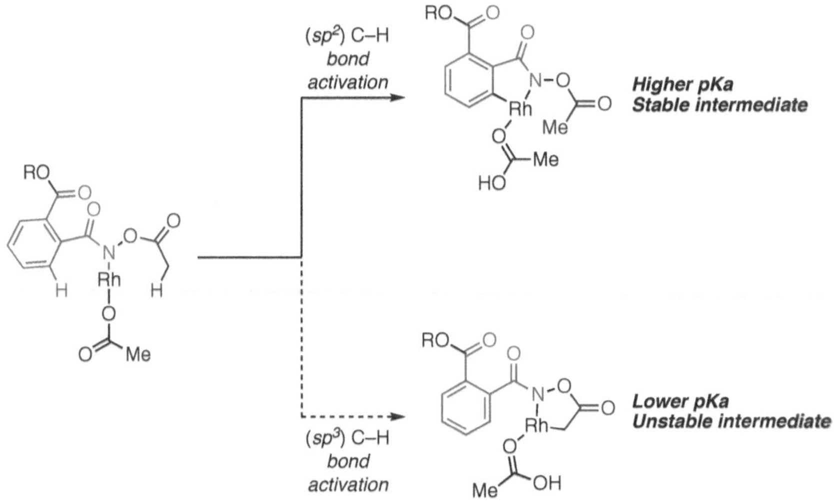

Figure 5.6 Proposed divergent C–H functionalization.

5.6 Activation of *N*-iminophthalimides

We next shifted our focus on the functionalization of *N*-iminoamines. Similar to **5-1a**, **5-5a** was subjected to alkene **5-2a** in the presence of a Cp*Rh(III)-precatalyst with base and methanol at reflux. We predicted that using the N–N bond as an internal oxidant could result in competing nitrogen sources for C–N bond formation. If the imine nitrogen is functionalized, cyclic carboamination would predominate, as opposed to acyclic carboamination by functionalization of the phthalimide nitrogen. Unfortunately, we did not see any desired product formation, but complete consumption of **5-5a**.

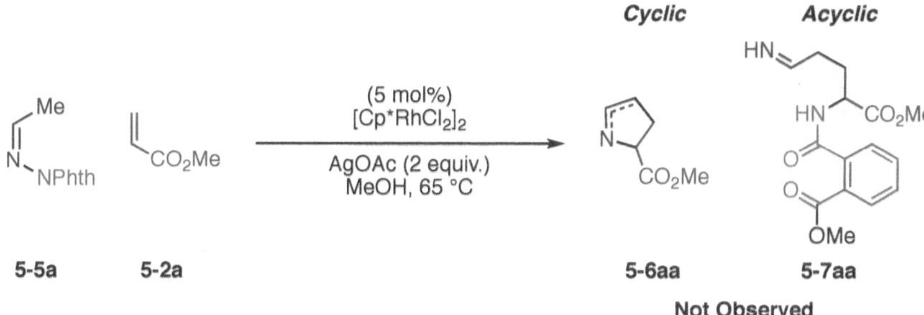

Figure 5.7 *Attempted carboamination of alkenes with N-iminophthalimides.*

Because **5-5a** was fully consumed, we decided to investigate the initial C–H activation using stoichiometric amounts of group 9 [Cp*MCl₂]₂ complexes. Begininng with cobalt, no desired metallacycle was observed for most likely 1 of 2 reasons–either it does not provide the desired C–H activation, or it does activate the C–H bond but is too unstable to isolate. Using rhodium, metallacycle **5-8ab** is formed in 73% yield. Gratifyingly, iridium also provides metallacycle **5-8ac** in moderate yield.

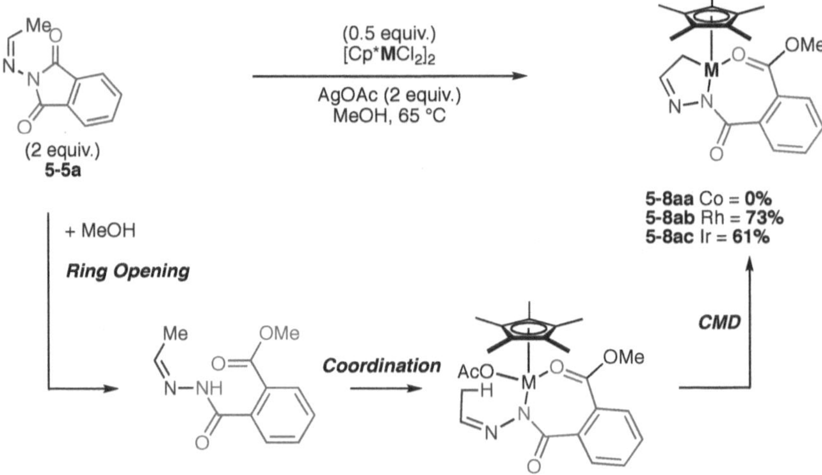

Figure 5.8 Isolation of metallacycles.

5.7 Future Directions for *sp³* C–H Activation *N*-iminophthalimides

With C–H activation experiments giving positive results, we now look to the future toward making this reactivity catalytic. Yu and coworkers recently reported using Boc-hydrazones in combination with internal alkynes under Rh(III)-catalysis the synthesis of 2,3,5-substituted pyrroles.[6] Importantly, the addition of AcOH provides a tautomerization from imine to enamine that sets up an *sp²* C–H activation. After migratory insertion of the alkyne and proposed metallacycle contraction, the C–N bond is formed by reductive elimination. Cleavage of the N–N bond after protonation turns over the catalyst.

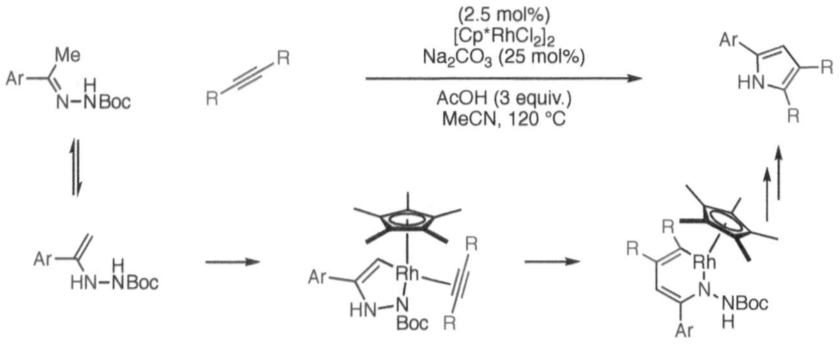

Figure 5.9 *Rh(III)-catalyzed pyrrole synthesis from Boc-hydrazones and alkynes.*

While this clearly provides a concise synthesis of substituted pyrroles, forcing conditions are still required. Potentially, the desired carboamination products could be observed with the addition of acid to our systems. Furthermore, other metals could provide a forward pathway for alkenes to render saturated N-heterocycles.

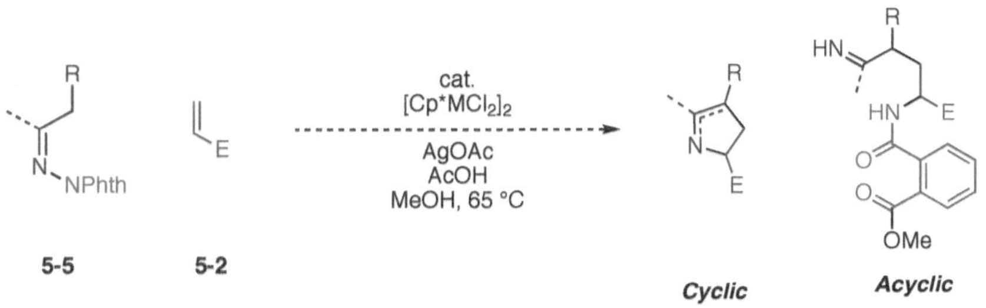

Figure 5.10 *Proposed cyclic and acyclic carboamination of N-iminophthalimides.*

5.8 Summary

Building on previous 1,2 carboamination success using *N*-enoxypthalimides, we investigated two viable alternatives that simplify pre-functionalization and take on the additional challenge of activating sp^3 C–H bonds. Using *N*-acetoxyphthlimides, sp^2 C–H bond activation predominates despite higher activation barriers. Alternative investigations with directing groups without competing sp^2 C–H bonds are underway. Finally, *N*-iminophthalimides show productive pathways toward sp^3 C–H bond activation. Isolation of Rh and Ir metallacycles show potential in furnishing new C–X bonds. Currently, stoichiometric studies are underway in an effort to push these substrates toward catalysis.

5.9 References

---

(1)   Piou, T; Rovis, T. *Nature* **2015**, *527*, 86.

(2)   Duchemin, C.; Cramer, N. *Angew. Chem. Int. Ed.* **2020**, *59*, 14129.

(3)   Zhang, Y.; Liu, H.; Tang, L.; Tang, H.-J.; Wang, L.; Zhu, C.; Feng, C. *J. Am. Chem. Soc.* **2018**, *140*, 10695.

(4)   Patil, A. S.; Mo, D.-L.; Wang, H.-Y.; Mueller, D. S.; Anderson, L. A. *Angew. Chem. Int. Ed.* **2012**, *51*, 7799.

(5)   Duchemin, C.; Cramer, N. *Org. Chem. Front.*, **2019**, *6*, 209.

(6)   Chan, C.-M.; Zhang, Z.; Yu, W.-Y. *Adv. Synth. Catal.* **2016**, *358*, 4067.

– Appendix A –

Supporting Information for Chapter Two

Rhodium(III)-Catalyzed Cyclopropanation of Unactivated Olefins Initiated by C–H Activation

Author: Erik J. T. Phipps, Tiffany Piou, Tomislav Rovis
Publication: Synlett
Publisher: Georg Thieme Verlag KG
Date: Jan 1, 2019

Review Order

Please review the order details and the associated terms and conditions.

No royalties will be charged for this reuse request although you are required to obtain a license and comply with the license terms and conditions. To obtain the license, click the Accept button below.

Licensed Content

Licensed Content Publisher	Georg Thieme Verlag KG
Licensed Content Publication	Synlett
Licensed Content Title	Rhodium(III)-Catalyzed Cyclopropanation of Unactivated Olefins Initiated by C–H Activation
Licensed Content Author	Erik J. T. Phipps, Tiffany Piou, Tomislav Rovis
Licensed Content Date	Jan 1, 2019
Licensed Content Volume	30
Licensed Content Issue	15

Order Details

Type of Use	Dissertation/Thesis
Requestor type	author of the original Thieme publication
Format	print and electronic
Portion	full article/document
Will you be translating?	no
Distribution quantity	1

About Your Work

Title	Rh(III)-catalyzed Difunctionalization of Alkenes Initiated by C–H Bond Activation
Institution name	Columbia University
Expected presentation date	Aug 2020

Additional Data

Portions	Modified figures and text from the article and SI to be used in the thesis.

Requestor Location

Mr. Erik Phipps
540 West 112th Street
Apt # 54

NEW YORK, NY 10025
United States
Attn: Columbia University Department of Chemistry

Tax Details

Publisher Tax ID	DE 147638607

Price

Total	0.00 USD

Rh(III)-Catalyzed Cyclopropanation of Unactivated Olefins Initiated by C–H Activation

Supporting Information

Erik J. T. Phipps, Tiffany Piou, and Tomislav Rovis*

Table of Contents

A1.1 General Methods

All reactions were carried out in oven-dried glassware with magnetic stirring. ACS grade TFE and reagents were purchased from TCI, Strem, Alfa Aesar, and Sigma-Aldrich and were used without further purification. Dichloromethane, tetrahydrofuran, diethyl ether were degassed with argon and passed through two columns of neutral alumina. Column chromatography was performed on SiliCycle® SilicaFlash® P60, 40-63 μm 60 Å and in general were run using flash techniques.[1] Thin layer chromatography was performed on SiliCycle® 250 μm 60 Å plates. Visualization was accomplished with UV light (254 nm). ^{1}H, ^{19}F, and ^{13}C NMR spectra were collected at ambient temperature in CDCl3 on Bruker 300 MHz, 400 MHz, or 500 MHz spectrometers. Chemical shifts are expressed as parts per million (δ, ppm) and are referenced to the residual solvent peak of chloroform (^{1}H = 7.26 ppm; ^{13}C = 77.2 ppm). Scalar coupling constants (J) are quoted in Hz. Multiplicity is reported as follows: s = singlet, d = doublet, t = triplet, q = quartet, m= multiplet). Mass spectra were obtained on a Waters Acquity PDA UPLC/MS (LRMS). Infrared (IR) spectra were obtained with neat samples on a Bruker Tensor 27 FT-IR spectrometer with OPUS software. Typically, the experiment consisted measuring the transmission in 8 scans in the region from 4000 to 400 cm^{-1}.

A1.2 Synthesis of Starting Materials

Synthesis of [Cp*CF3RhCl2]2 Catalyst[2, 3]

Synthesis of 1,2,3,4-tetramethyl-5-(trifluoromethyl)cyclopenta-1,3-diene (+ isomers)

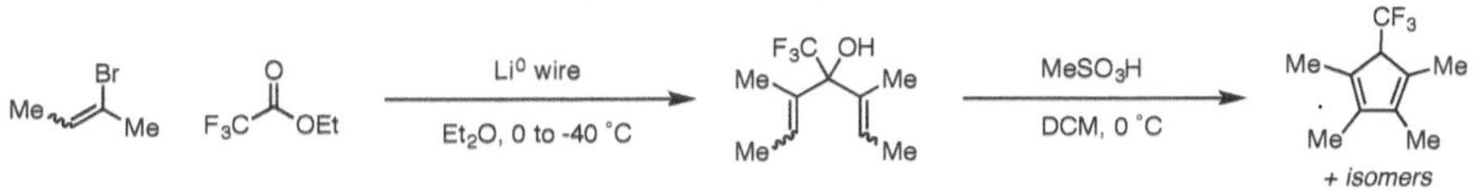

Figure 1.

This procedure was performed according to literature precedent.

Synthesis of [Cp*CF3RhCl2]2

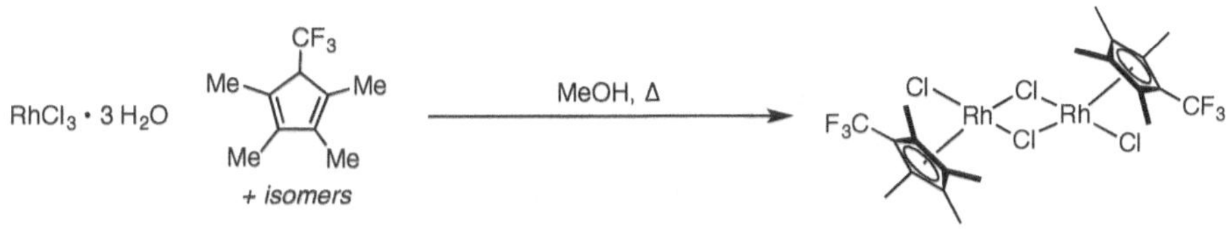

Figure 2.

This procedure was performed according to literature precedent.

Synthesis of *N*-enoxyphthalimides

Method A[4] (**1a-1i**)

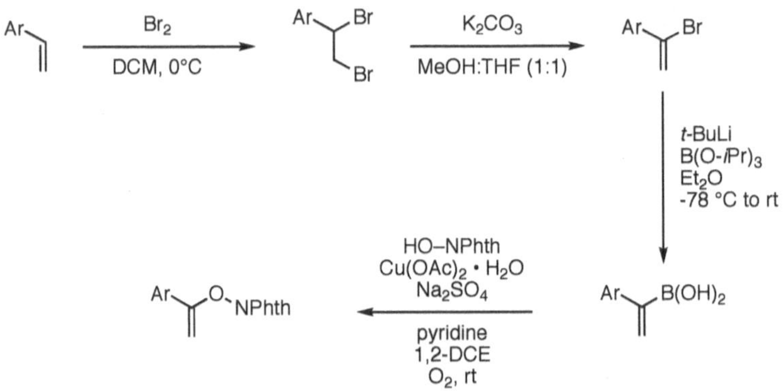

Figure 3.

This procedure was performed according to literature precedent.

Method B[5] (**1j**)

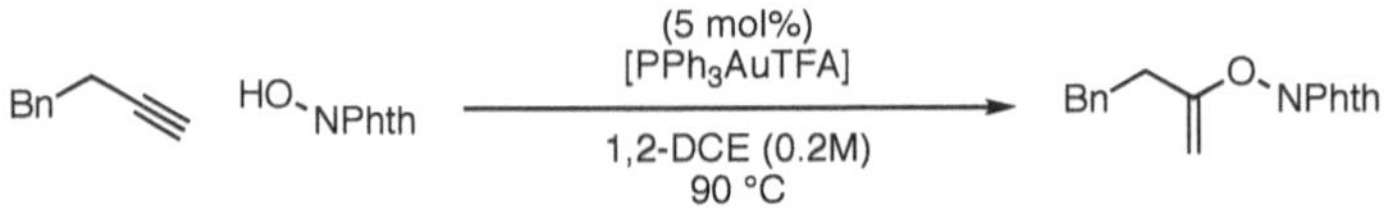

Figure 4.

This procedure was performed according to literature precedent.

Synthesis of alkene coupling partners[6] (**2e-2i**)

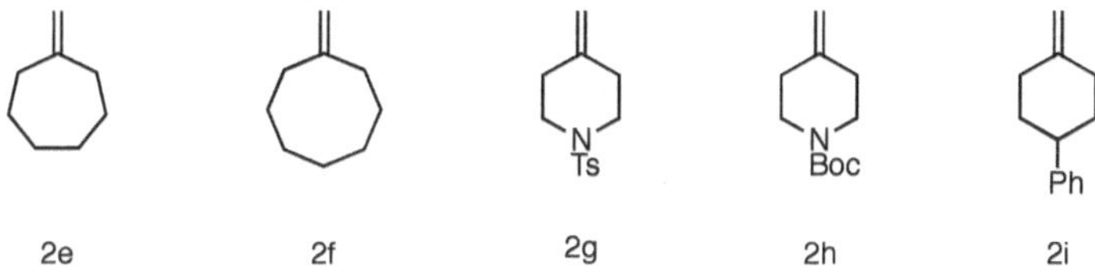

Figure 5

This procedure was performed according to literature precedent.[7]

2e   2f   2g   2h   2i

A1.3 General Procedure for the Cyclopropanation Reaction and Characterization of Products

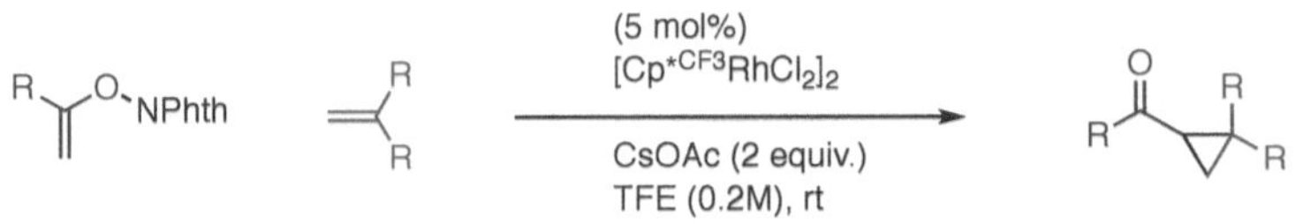

Figure 6.

N-enoxyphthalimide (0.1 mmol), catalyst [Cp*^{CF3}RhCl$_2$]$_2$ (5 mol%, 0.005 mmol, 3.7 mg), and CsOAc (2 equiv., 0.2 mmol, 38.5 mg) were weighed in a 1-dram vial with a magnetic stirbar. TFE (0.2 M, 500 μL) was added followed by alkene (1.2 equiv., 0.12 mmol). The vial was sealed with a screw-cap and stirred at room temperature for 12 hours. Upon completion judged by TLC, the crude solution was diluted with EtOAc and partitioned with the addition of DI water. The aqueous layer was extracted three times with EtOAc and the combined organic extracts were filtered through a pad of celite® and Na$_2$SO$_4$ then concentrated. The crude residue was purified by flash chromatography (Hexane:EtOAc, 19:1) to afford the cyclopropane product.

Reaction Optimization

Ph–C(=O)–O–NPhth + 2-methylene hexane (Me, *n*-Bu) (1.2 equiv.) → [Cp*^{CF3}RhCl$_2$]$_2$ (5 mol%), Base (2 equiv.), TFE (0.2M), rt → Ph–C(=O)–cyclopropane(Me, *n*-Bu)

1:1 d.r.

Entry	Base	Yield
1	KOPiv	59%
2	KOAc	76%
3	1-AdCO$_2$K	66%
4	LiOAc	64%
5	NaOAc	73%
6	CsOAc	82%

We first examined carboxylate bases beginning with the standard conditions from reference 3 using 2-methylenehexane to afford the two diastereomers of cyclopropane product. When KOAc proved to be the best we moved on to testing different alkali metal cations. We decided to move forward using CsOAc as our base.

Alkene	Product	NMR Yield	d.r.
(2-methylenehexane; Me, *n*-Bu)	(cyclopropyl ketone; Ph, Me, *n*-Bu)	82%	1:1
(Me, *i*-Pr)	(cyclopropyl ketone; Ph, Me, *i*-Pr)	17%	1:1
(H, *n*-Oct)	(cyclopropyl ketone; Ph, *n*-Oct)	58%	1:1
(methylenecyclohexane)	(spiro cyclopropyl ketone; Ph)	99%	--

Additionally, we surveyed different alkenes as potential coupling partners. From our previous optimization screen, we observed asymmetric 1,1-disubstituted olefins are not selective for one diastereomer but the yield drastically drops when the steric load is increased from 2-methylenhexane to 2,3-dimethylbut-1-ene. Using 1-decene, we observed moderate yield for the cyclopropane; however, the reaction remained unselective. Finally, we considered symmetrical 1,1-disubstituted alkenes where we observed methylencyclohexane gives 99% yield. While these products do not provide access to a single diastereomer, we considered this method could improve the synthesis of interesting spirocyclic species that can be difficult to access.

Characterization of Products

3aa (2,2-diethylcyclopropyl)(phenyl)methanone

Chemical Formula: $C_{14}H_{18}O$
Exact Mass: 202.14

Yield = 40% Colorless oil. R_f = 0.74 (4:1 hexanes: Ethyl Acetate).

¹H NMR (500 MHz, Chloroform-*d*) δ 7.99 (dt, *J* = 7.2, 1.3 Hz, 2H), 7.59 – 7.51 (m, 1H), 7.46 (t, *J* = 7.7 Hz, 2H), 2.51 (dd, *J* = 7.4, 5.6 Hz, 1H), 1.72 (dq, *J* = 14.7, 7.4 Hz, 1H), 1.56 – 1.37 (m, 4H), 1.02 (t, *J* = 7.4 Hz, 3H), 0.96 (dd, *J* = 7.5, 4.0 Hz, 1H), 0.77 (t, *J* = 7.4 Hz, 3H).

¹³C NMR (126 MHz, CDCl₃) δ 198.7, 139.2, 132.5, 128.6, 128.2, 38.3, 32.1, 29.6, 21.3, 21.1, 11.3, 10.8.

IR(neat): 2963, 2930, 1666, 1448, 1396, 1215, 1024, 982, 712, 689 cm⁻¹

LRMS m/z (ESI APCI): calculated for $C_{14}H_{18}O$ [M+H] 203.1, found 203.1.

3ab 1-benzoylspiro[2.3]hexane-5,5-dicarbonitrile

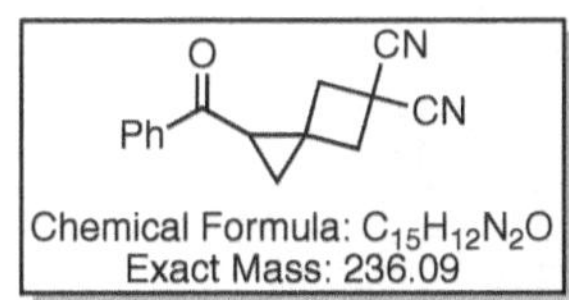

Yield = 56% Colorless oil. R_f = 0.39 (4:1 hexanes: Ethyl Acetate).

¹H NMR (500 MHz, Chloroform-*d*) δ 7.98 (dt, *J* = 8.5, 1.5 Hz, 2H), 7.60 (td, *J* = 7.3, 1.5 Hz, 1H), 7.50 (td, *J* = 7.7, 1.6 Hz, 2H), 3.15 (dtdd, *J* = 9.6, 8.1, 6.6, 1.6 Hz, 1H), 2.76 (ddd, *J* = 7.6, 5.6, 1.4 Hz, 1H), 2.69 (qd, *J* = 6.8, 5.8, 1.5 Hz, 2H), 2.54 (ddd, *J* = 13.0, 9.5, 1.5 Hz, 1H), 1.61 (td, *J* = 5.3, 1.5 Hz, 1H), 1.33 (ddd, *J* = 8.4, 4.8, 1.5 Hz, 1H).

¹³C NMR (126 MHz, CDCl₃) δ 197.6, 138.2, 133.3, 128.9, 128.1, 122.6, 35.2, 32.2, 31.5, 29.1, 22.3, 18.1.

IR(neat): 2992, 2942, 2236, 1661, 1449, 1390, 1335, 1230, 1012, 715, 689 cm⁻¹

LRMS m/z (ESI APCI): calculated for $C_{15}H_{12}N_2O$ [M+H] 237.1, found 237.1.

3ac phenyl(spiro[2.4]heptan-1-yl)methanone

Chemical Formula: $C_{14}H_{16}O$
Exact Mass: 200.12

Yield = 87% Colorless oil. R_f = 0.78 (4:1 hexanes: Ethyl Acetate).

^{1}H NMR (500 MHz, Chloroform-*d*) δ 8.02 – 7.90 (m, 2H), 7.59 – 7.51 (m, 1H), 7.51 – 7.42 (m, 2H), 2.69 (dd, *J* = 7.6, 5.5 Hz, 1H), 1.89 – 1.80 (m, 1H), 1.79 – 1.50 (m, 8H), 1.19 (dd, *J* = 7.7, 3.9 Hz, 1H).

^{13}C NMR (126 MHz, CDCl$_3$) δ 199.1, 139.1, 132.6, 128.6, 128.0, 38.8, 37.5, 32.6, 30.0, 26.3, 26.2, 22.1.

IR(neat): 2952, 2864, 1665, 1448, 1390, 1216, 1012, 715, 691 cm^{-1}

LRMS m/z (ESI APCI): calculated for $C_{14}H_{16}O$ [M+H] 201.1, found 201.1.

3ad phenyl(spiro[2.5]octan-1-yl)methanone

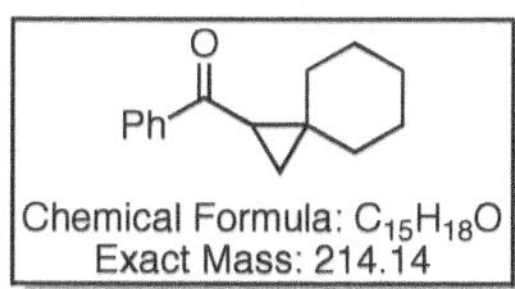

Chemical Formula: $C_{15}H_{18}O$
Exact Mass: 214.14

Yield = 98% Colorless oil. R_f = 0.73 (4:1 hexanes: Ethyl Acetate).

¹H NMR (500 MHz, Chloroform-*d*) δ 8.06 – 7.97 (m, 2H), 7.59 – 7.51 (m, 1H), 7.51 – 7.43 (m, 2H), 2.51 (dd, *J* = 7.3, 5.4 Hz, 1H), 1.70 – 1.39 (m, 10H), 1.19 (dt, *J* = 12.6, 6.1 Hz, 1H), 0.95 (dd, *J* = 7.4, 4.0 Hz, 1H).

¹³C NMR (126 MHz, $CDCl_3$) δ 198.4, 139.1, 132.5, 128.6, 128.2, 38.0, 35.6, 32.2, 28.48, 26.3, 26.2, 26.0, 21.5.

IR(neat): 2921, 2850, 1664, 1447, 1396, 1216, 980, 718, 689 cm⁻¹

LRMS m/z (ESI APCI): calculated for $C_{15}H_{18}O$ [M+H] 215.1, found 215.1.

3ae phenyl(spiro[2.6]nonan-1-yl)methanone

Chemical Formula: $C_{16}H_{20}O$
Exact Mass: 228.15

Yield = 70% Colorless oil. R_f = 0.74 (4:1 hexanes: Ethyl Acetate).

^{1}H NMR (500 MHz, Chloroform-*d*) δ 8.07 – 7.97 (m, 2H), 7.60 – 7.52 (m, 1H), 7.47 (dd, *J* = 8.3, 6.9 Hz, 2H), 2.54 (dd, *J* = 7.5, 5.7 Hz, 1H), 1.80 – 1.66 (m, 4H), 1.58 (tdd, *J* = 14.0, 5.9, 3.9 Hz, 6H), 1.49 (ddt, *J* = 10.9, 7.4, 5.5 Hz, 2H), 1.32 (dddd, *J* = 15.9, 9.5, 7.4, 3.6 Hz, 1H), 0.99 (dd, *J* = 7.5, 3.9 Hz, 1H).

^{13}C NMR (126 MHz, CDCl$_3$) δ 198.7, 139.2, 132.6, 128.6, 128.2, 40.8, 37.1, 33.5, 30.53, 28.2, 28.1, 26.6, 26.5, 23.3.

IR(neat): 2920, 2852, 1665, 1448, 1395, 1217, 981, 710, 689 cm[-1]

LRMS m/z (ESI APCI): calculated for $C_{16}H_{20}O$ [M+H] 229.2, found 229.2.

3af phenyl(spiro[2.7]decan-1-yl)methanone

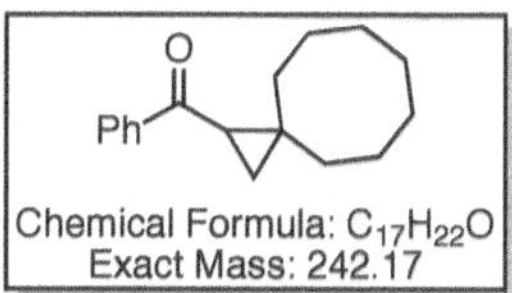

Yield = 53% Colorless oil. R$_f$ = 0.69 (4:1 hexanes: Ethyl Acetate).

^{1}H NMR (500 MHz, Chloroform-*d*) δ 8.04 – 7.92 (m, 2H), 7.58 – 7.52 (m, 1H), 7.47 (dd, *J* = 8.4, 6.9 Hz, 2H), 2.50 (dd, *J* = 7.5, 5.7 Hz, 1H), 1.92 (ddd, *J* = 14.4, 8.7, 2.8 Hz, 1H), 1.82 – 1.56 (m, 10H), 1.55 – 1.38 (m, 4H), 1.00 (dd, *J* = 7.5, 4.0 Hz, 1H).

^{13}C NMR (126 MHz, CDCl$_3$) δ 198.7, 139.3, 132.6, 128.6, 128.2, 39.1, 36.6, 34.6, 27.9, 27.4, 26.9, 25.9, 25.6, 25.3, 23.3.

IR(neat): 2944, 2911, 1668, 1472, 1447, 1216, 1010, 945, 734, 702 cm^{-1}

LRMS m/z (ESI APCI): calculated for C$_{21}$H$_{22}$O [M+H] 243.2, found 243.2.

3ag phenyl(6-tosyl-6-azaspiro[2.5]octan-1-yl)methanone

O
Ph
NTs
Chemical Formula: C$_{21}$H$_{23}$NO$_3$S
Exact Mass: 369.14

Yield = 72% White solid. R$_f$ = 0.22 (4:1 hexanes: Ethyl Acetate).

^{1}H NMR (500 MHz, Chloroform-*d*) δ 7.82 (dd, *J* = 8.3, 1.4 Hz, 2H), 7.61 – 7.56 (m, 2H), 7.54 – 7.49 (m, 1H), 7.40 – 7.33 (m, 2H), 7.27 – 7.24 (m, 2H), 3.14 (qdd, *J* = 11.4, 7.0, 4.6 Hz, 2H), 2.94 (ddd, *J* = 11.3, 7.2, 3.8 Hz, 1H), 2.76 (ddd, *J* = 11.7, 7.3, 3.7 Hz, 1H), 2.50 (dd, *J* = 7.6, 5.5 Hz, 1H), 2.42 (s, 3H), 1.73 (dtdd, *J* = 21.2, 17.2, 13.1, 8.8 Hz, 4H), 1.51 (dd, *J* = 5.4, 4.5 Hz, 1H), 0.95 (dd, *J* = 7.6, 4.4 Hz, 1H).

^{13}C NMR (126 MHz, CDCl$_3$) δ 197.5, 143.6, 138.5, 133.0, 132.8, 129.7, 128.7, 128.0, 127.8, 46.5, 46.1, 36.1, 31.7, 30.6, 27.5, 21.7, 20.3.

IR(neat): 2972, 2819, 1666, 1602, 1451, 1373, 1239, 1170, 890, 711, 688 cm^{-1}

LRMS m/z (ESI APCI): calculated for C$_{21}$H$_{23}$ NO$_3$S [M+H] 370.1, found 370.1.

3ah *tert*-butyl 1-benzoyl-6-azaspiro[2.5]octane-6-carboxylate

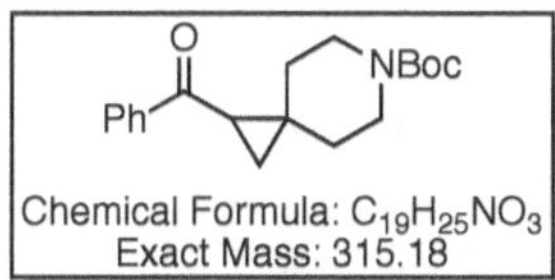

Yield = 84% Colorless oil. R_f = 0.45 (4:1 hexanes: Ethyl Acetate).

¹H NMR (500 MHz, Chloroform-*d*) δ 8.02 – 7.96 (m, 2H), 7.60 – 7.53 (m, 1H), 7.47 (dd, *J* = 8.3, 6.9 Hz, 2H), 3.61 – 3.49 (m, 2H), 3.37 (ddd, *J* = 13.1, 6.6, 4.2 Hz, 1H), 3.13 (ddd, *J* = 13.2, 7.0, 4.3 Hz, 1H), 2.62 (dd, *J* = 7.5, 5.4 Hz, 1H), 1.64 – 1.57 (m, 4H), 1.43 (s, 9H), 1.05 (dd, *J* = 7.5, 4.2 Hz, 1H).

¹³C NMR (126 MHz, CDCl₃) δ 197.8, 155.0, 138.7, 134.5, 132.9, 128.8, 128.2, 123.8, 79.7, 77.4, 36.8, 33.0, 30.9, 28.6, 20.6.

IR(neat): 2975, 2818, 1666, 1419, 1365, 1238, 1166, 1120, 903, 720, 689 cm⁻¹

LRMS m/z (ESI APCI): calculated for C₁₉H₂₅NO₃ [M+H] 316.2, found 316.2.

3ai phenyl(6-phenylspiro[2.5]octan-1-yl)methanone

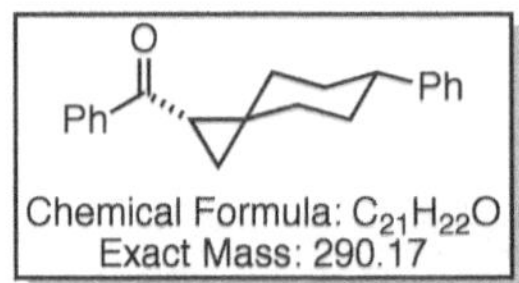

Chemical Formula: $C_{21}H_{22}O$
Exact Mass: 290.17

Yield = 97% White solid. R_f = 0.58 (4:1 hexanes: Ethyl Acetate).

^{1}H NMR (500 MHz, Chloroform-*d*) δ 8.02 – 7.94 (m, 2H), 7.61 – 7.54 (m, 1H), 7.53 – 7.46 (m, 2H), 7.34 – 7.28 (m, 2H), 7.27 – 7.23 (m, 2H), 7.23 – 7.17 (m, 1H), 2.58 (dq, *J* = 8.5, 5.6, 4.6 Hz, 2H), 2.25 (tdd, *J* = 12.9, 3.8, 1.7 Hz, 1H), 1.94 (ddt, *J* = 21.9, 12.7, 2.7 Hz, 2H), 1.77 – 1.56 (m, 4H), 1.48 (tdd, *J* = 12.9, 3.7, 1.5 Hz, 1H), 1.31 (dq, *J* = 13.0, 3.0 Hz, 1H), 0.99 (ddd, *J* = 7.4, 4.0, 1.6 Hz, 1H).

^{13}C NMR (126 MHz, CDCl$_3$) δ 198.2, 147.1, 139.1, 132.7, 128.7, 128.6, 128.2, 127.0, 126.2, 44.4, 40.3, 37.7, 33.6, 33.5, 32.4, 29.0, 20.4.

IR(neat): 2921, 2872, 1671, 1492, 1277, 1216, 970, 755, 698 cm^{-1}

LRMS m/z (ESI APCI): calculated for $C_{21}H_{22}O$ [M+H] 291.2, found 291.2.

3bd spiro[2.5]octan-1-yl(*p*-tolyl)methanone

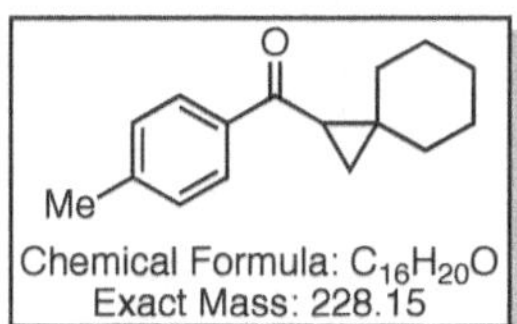

Yield = 97% Colorless oil. R_f = 0.81 (4:1 hexanes: Ethyl Acetate).

^{1}H NMR (500 MHz, Chloroform-*d*) δ 7.92 (d, *J* = 8.2 Hz, 2H), 7.26 (d, *J* = 7.9 Hz, 2H), 2.49 (dd, *J* = 7.3, 5.4 Hz, 1H), 2.42 (s, 3H), 1.73 – 1.54 (m, 4H), 1.47 (dddd, *J* = 28.0, 16.4, 9.6, 4.9 Hz, 6H), 1.17 (dt, *J* = 12.4, 6.1 Hz, 1H), 0.92 (dd, *J* = 7.3, 4.0 Hz, 1H).

^{13}C NMR (126 MHz, CDCl₃) δ 197.9, 143.2, 136.7, 129.3, 128.3, 38.0, 35.2, 32.1, 28.5, 26.4, 26.2, 26.0, 21.8, 21.2.

IR(neat): 2923, 2854, 1663, 1410, 1217, 1155, 854, 806, 598 cm⁻¹

LRMS m/z (ESI APCI): calculated for C₁₆H₂₀O [M+H] 229.2, found 229.2.

3cd (4-(*tert*-butyl)phenyl)(spiro[2.5]octan-1-yl)methanone

Chemical Formula: $C_{19}H_{26}O$
Exact Mass: 270.20

Yield = 89% Colorless oil. R_f = 0.75 (4:1 hexanes: Ethyl Acetate).

^{1}H NMR (500 MHz, Chloroform-*d*) δ 7.99 – 7.93 (m, 2H), 7.51 – 7.44 (m, 2H), 2.49 (dd, *J* = 7.4, 5.4 Hz, 1H), 1.70 – 1.39 (m, 10H), 1.35 (s, 9H), 1.21 (dd, *J* = 11.2, 5.8 Hz, 1H), 0.92 (dd, *J* = 7.4, 4.0 Hz, 1H).

^{13}C NMR (126 MHz, CDCl$_3$) δ 198.0, 156.1, 136.6, 128.2, 125.6, 38.0, 35.2, 32.1, 31.3, 28.5, 26.4, 26.2, 26.1, 21.2.

IR(neat): 2903, 2868, 1662, 1409, 1223, 854, 808, 598 cm^{-1}

LRMS m/z (ESI APCI): calculated for $C_{19}H_{26}O$ [M+H] 271.2, found 271.2.

3dd (4-fluorophenyl)(spiro[2.5]octan-1-yl)methanone

Chemical Formula: $C_{15}H_{17}FO$
Exact Mass: 232.13

Yield = 96% Colorless oil. R_f = 0.75 (4:1 hexanes: Ethyl Acetate).

¹H NMR (500 MHz, Chloroform-*d*) δ 8.08 – 7.99 (m, 2H), 7.19 – 7.10 (m, 2H), 2.45 (dd, *J* = 7.4, 5.4 Hz, 1H), 1.69 – 1.56 (m, 4H), 1.54 – 1.41 (m, 6H), 1.15 (d, *J* = 15.6 Hz, 1H), 0.95 (dd, *J* = 7.3, 4.0 Hz, 1H).

¹³C NMR (126 MHz, CDCl₃) δ 196.7, 165.5 (d, *J* = 253.8 Hz), 135.5 (d, *J* = 3.0 Hz), 130.7 (d, *J* = 9.1 Hz), 115.7 (d, *J* = 21.6 Hz), 38.0, 35.7, 32.1, 28.5, 26.3, 26.2, 26.0, 21.5.

IR(neat): 2929, 2854, 1665, 1506, 1216, 1155, 854, 710, 598 cm⁻¹

LRMS m/z (ESI APCI): calculated for $C_{15}H_{17}FO$ [M+H] 233.1, found 233.1.

3ed spiro[2.5]octan-1-yl(*m*-tolyl)methanone

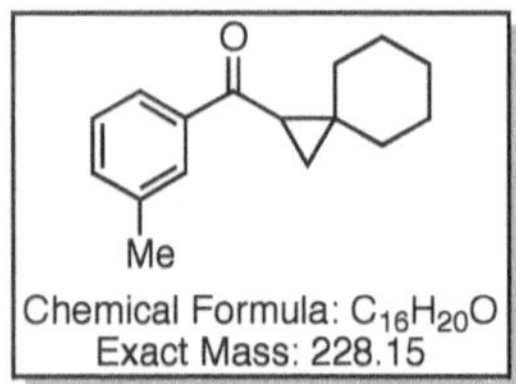

Yield = 75% Colorless oil. R_f = 0.84 (4:1 hexanes: Ethyl Acetate).

¹H NMR (500 MHz, Chloroform-*d*) δ 7.84 – 7.79 (m, 2H), 7.36 (m, 1H), 7.35 (m, 1H), 2.50 (dd, *J* = 7.4, 5.5 Hz, 1H), 2.42 (s, 3H), 1.70 – 1.40 (m, 10H), 1.20 (dt, *J* = 12.1, 6.0 Hz, 1H), 0.93 (dd, *J* = 7.3, 4.0 Hz, 1H).

¹³C NMR (126 MHz, CDCl₃) δ 198.5, 139.2, 138.4, 133.3, 128.7, 128.5, 125.5, 38.0, 35.5, 32.2, 28.5, 26.3, 26.2, 26.0, 21.6, 21.5.

IR(neat): 2922, 2857, 1664, 1240, 1180, 755, 688 cm⁻¹

LRMS m/z (ESI APCI): calculated for C₁₆H₂₀O [M+H] 229.2, found 229.2.

3fd (3-methoxyphenyl)(spiro[2.5]octan-1-yl)methanone

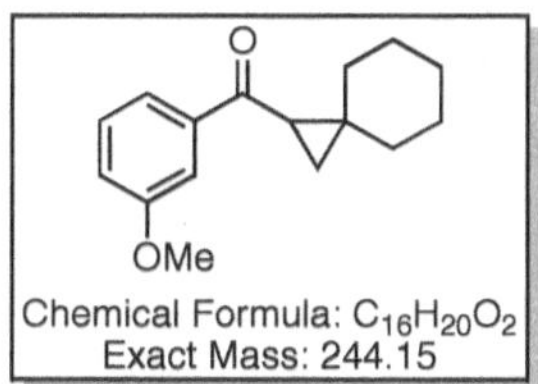

Yield = 67% Colorless oil. R_f = 0.72 (4:1 hexanes: Ethyl Acetate).

¹H NMR (500 MHz, Chloroform-*d*) δ 7.62 (dd, *J* = 7.7, 1.5 Hz, 1H), 7.52 (t, *J* = 2.0 Hz, 1H), 7.37 (t, *J* = 7.9 Hz, 1H), 7.09 (dd, *J* = 8.2, 2.6 Hz, 1H), 3.86 (s, 3H), 2.49 (dd, *J* = 7.3, 5.5 Hz, 1H), 1.70 – 1.40 (m, 8H), 1.19 (m, 1H), 0.94 (dd, *J* = 7.3, 4.0 Hz, 1H).

¹³C NMR (126 MHz, CDCl₃) δ 198.1, 159.9, 140.5, 129.6, 120.9, 118.9, 112.5, 55.6, 38.0, 35.7, 32.3, 28.5, 26.3, 26.1, 26.0, 21.6.

IR(neat): 2921, 2850, 1664, 1595, 1448, 1284, 1259, 1034, 870, 845, 762, 684 cm⁻¹

LRMS m/z (ESI APCI): calculated for $C_{16}H_{20}O_2$ [M+H] 245.2, found 245.2.

3gd (3-fluorophenyl)(spiro[2.5]octan-1-yl)methanone

Chemical Formula: $C_{15}H_{17}FO$
Exact Mass: 232.13

Yield = 90% Colorless oil. R_f = 0.75 (4:1 hexanes: Ethyl Acetate).

^{1}H NMR (500 MHz, Chloroform-*d*) δ 7.80 (dt, *J* = 7.8, 1.3 Hz, 1H), 7.67 (ddd, *J* = 9.6, 2.7, 1.6 Hz, 1H), 7.44 (td, *J* = 8.0, 5.5 Hz, 1H), 7.28 – 7.20 (m, 1H), 2.45 (dd, *J* = 7.4, 5.4 Hz, 1H), 1.70 – 1.39 (m, 10H), 1.17 (d, *J* = 9.8 Hz, 1H), 0.98 (dd, *J* = 7.3, 4.0 Hz, 1H).

^{13}C NMR (126 MHz, $CDCl_3$) δ 197.0 (d, *J* = 2.1 Hz), 163.0 (d, *J* = 247.5 Hz), 141.2 (d, *J* = 6.2 Hz), 130.3 (d, *J* = 7.7 Hz), 123.9 (d, *J* = 3.2 Hz), 119.5 (d, *J* = 21.6 Hz), 115.0 (d, *J* = 22.2 Hz), 38.0, 36.2, 32.3, 28.4, 26.3, 26.2, 26.1, 21.9.

IR(neat): 2929, 2854, 1667, 1511, 1410, 1214, 1111, 830, 757 cm[-1]

LRMS m/z (ESI APCI): calculated for $C_{15}H_{17}FO$ [M+H] 233.1, found 233.1.

3hd (2-fluorophenyl)(spiro[2.5]octan-1-yl)methanone

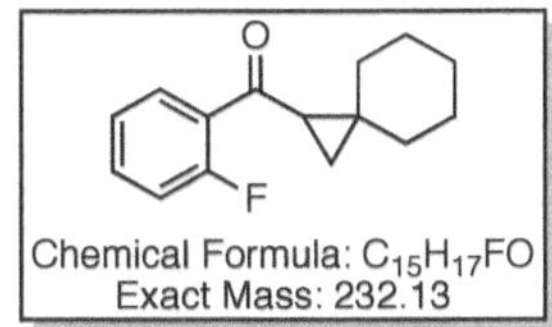

Yield = 59% Colorless oil. R_f = 0.75 (4:1 hexanes: Ethyl Acetate).

¹H NMR (500 MHz, Chloroform-*d*) δ 7.76 (td, *J* = 7.6, 1.9 Hz, 1H), 7.48 (tdd, *J* = 7.2, 4.9, 1.9 Hz, 1H), 7.21 (t, *J* = 7.5 Hz, 1H), 7.13 (dd, *J* = 10.9, 8.4 Hz, 1H), 2.48 (ddd, *J* = 7.2, 5.5, 3.4 Hz, 1H), 1.54 (ddtd, *J* = 33.0, 28.4, 11.4, 10.3, 5.2 Hz, 10H), 1.32 – 1.21 (m, 1H), 0.95 (dd, *J* = 7.3, 3.9 Hz, 1H).

¹³C NMR (126 MHz, Chloroform-*d*) δ 197.0 (d, *J* = 3.2 Hz), 161.6 (d, *J* = 253.5 Hz), 133.8 (d, *J* = 8.7 Hz), 130.5 (d, *J* = 2.7 Hz), 128.3 (d, *J* = 13.1 Hz), 124.4 (d, *J* = 3.5 Hz), 116.7 (d, *J* = 23.5 Hz), 37.9, 36.9, 36.3, 36.3, 28.4, 26.3, 26.2, 25.8, 25.8, 22.6.

IR(neat): 2930, 2854, 1672, 1479, 1453, 1346, 1204, 1102, 829, 757 cm⁻¹

LRMS m/z (ESI APCI): calculated for C₁₅H₁₇FO [M+H] 233.1, found 233.1.

3id naphthalen-2-yl(spiro[2.5]octan-1-yl)methanone

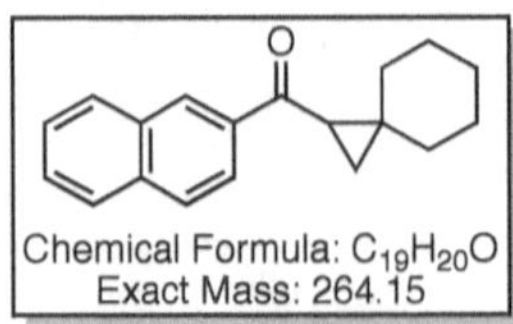

Yield = 67% Colorless oil. R$_f$ = 0.77 (4:1 hexanes: Ethyl Acetate).

¹**H NMR** (500 MHz, Chloroform-*d*) δ 8.56 (d, *J* = 1.7 Hz, 1H), 8.08 (dd, *J* = 8.6, 1.8 Hz, 1H), 8.04 – 7.96 (m, 1H), 7.90 (dd, *J* = 10.1, 8.1 Hz, 2H), 7.57 (dddd, *J* = 18.9, 8.1, 6.8, 1.4 Hz, 2H), 2.67 (dd, *J* = 7.4, 5.4 Hz, 1H), 1.57 (tddd, *J* = 50.6, 21.1, 10.8, 5.2 Hz, 10H), 1.20 (dq, *J* = 11.9, 6.1, 5.0 Hz, 1H), 1.00 (dd, *J* = 7.3, 4.0 Hz, 1H).

¹³**C NMR** (126 MHz, CDCl$_3$) δ 198.1, 136.5, 135.5, 132.8, 129.7, 129.6, 128.4, 128.3, 127.9, 126.8, 124.4, 38.1, 35.7, 32.3, 28.6, 26.3, 26.2, 26.1, 21.6.

IR(neat): 2916, 2846, 1654, 1398, 1183, 1125, 1116, 808, 780 cm⁻¹

LRMS m/z (ESI APCI): calculated for C$_{19}$H$_{20}$O [M+H] 265.2, found 265.2.

3jd 3-phenyl-1-(spiro[2.5]octan-1-yl)propan-1-one

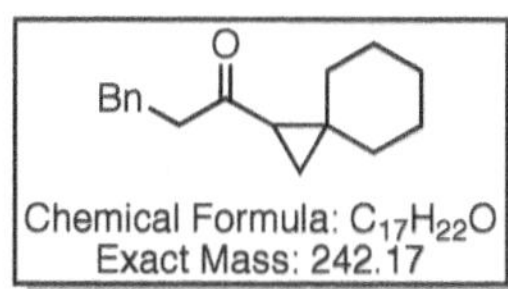

Yield = 98% Colorless oil. R$_f$ = 0.81 (4:1 hexanes: Ethyl Acetate).

¹H NMR (500 MHz, Chloroform-*d*) δ 7.31 – 7.26 (m, 2H), 7.24 – 7.17 (m, 3H), 2.96 – 2.88 (m, 4H), 1.79 (dd, *J* = 7.4, 5.4 Hz, 1H), 1.58 – 1.38 (m, 9H), 1.28 (dd, *J* = 5.5, 3.9 Hz, 1H), 1.13 (p, *J* = 7.5 Hz, 1H), 0.80 (dd, *J* = 7.4, 4.0 Hz, 1H).

¹³C NMR (126 MHz, CDCl$_3$) δ 207.7, 141.5, 128.6, 128.5, 126.2, 46.4, 37.9, 35.2, 34.7, 30.3, 28.1, 26.3, 26.2, 26.1, 22.1.

IR(neat): 2921, 2850, 1692, 1445, 1398, 1117, 1082, 748, 699 cm^{-1}

LRMS m/z (ESI APCI): calculated for C$_{17}$H$_{22}$O [M+H] 243.2, found 243.2.

A1.4 Mechanistic Experiments

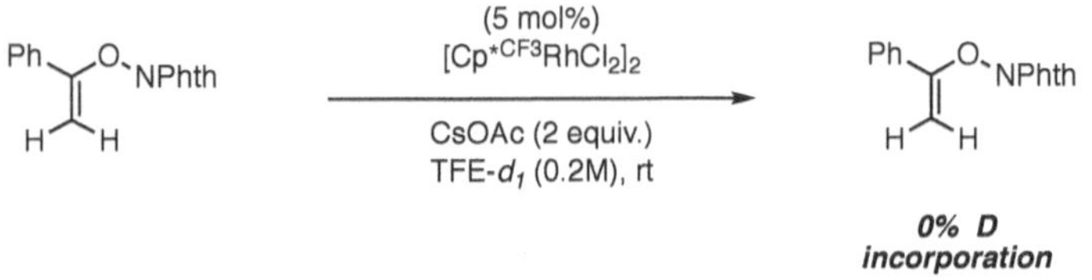

Figure 7.

N-enoxyphthalimide **1a** (0.1 mmol), catalyst [Cp*CF3RhCl2]2 (5 mol%, 0.005 mmol), and CsOAc (2 equiv., 0.2 mmol) were weighed in a 1-dram vial with a magnetic stirbar. TFE-d_1 (0.2 M, 500 µL) was added. The vial was sealed with a screw-cap and stirred for 3 hours. Upon completion judged by TLC, the crude solution was diluted with EtOAc and partitioned with the addition of DI water. The aqueous layer was extracted three times with EtOAc and the combined organic extracts were filtered through a pad of celite® and Na2SO4 then concentrated. The crude residue was purified by flash chromatography (Hexane:EtOAc, 19:1) to afford the starting material.

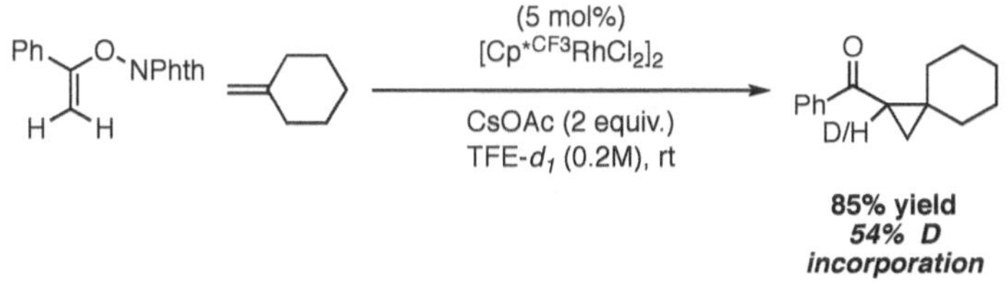

Figure 8.

N-enoxyphthalimide **1a** (0.1 mmol), catalyst [Cp*^{CF3}RhCl$_2$]$_2$ (5 mol%, 0.005 mmol), and CsOAc (2 equiv., 0.2 mmol) were weighed in a 1-dram vial with a magnetic stirbar. TFE (0.2 M, 500 μL) was added followed by addition of alkene **2d** (1.2 equiv. 0.12 mmol). The vial was sealed with a screw-cap and stirred for 3 hours. Upon completion judged by TLC, the crude solution was diluted with EtOAc and partitioned with the addition of DI water. The aqueous layer was extracted three times with EtOAc and the combined organic extracts were filtered through a pad of celite® and Na$_2$SO$_4$ then concentrated. The crude residue was purified by flash chromatography (Hexane:EtOAc, 19:1) to afford cyclopropane **3ad'**.

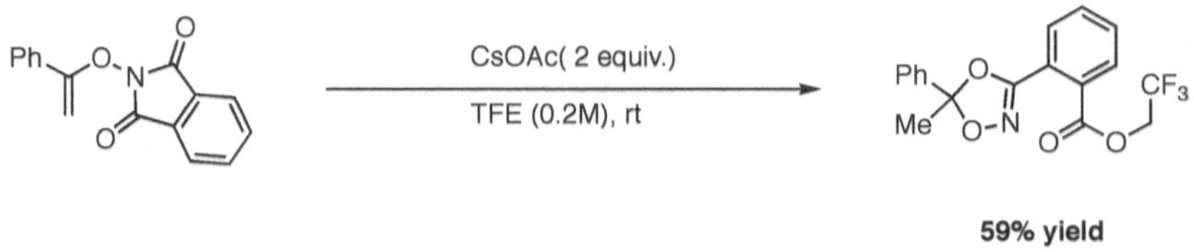

Figure 9.

N-enoxyphthalimide **1a** (0.1 mmol) and CsOAc (2 equiv., 0.2 mmol, 38.5 mg) were weighed in a 1-dram vial with a magnetic stirbar. TFE (0.2 M, 500 μL) was added and the vial was sealed with a screw-cap and stirred at room temperature for 12 hours. Upon completion judged by TLC, the crude solution was diluted with EtOAc and partitioned with the addition of DI water. The aqueous layer was extracted three times with EtOAc and the combined organic extracts were filtered through a pad of celite® and Na_2SO_4 then concentrated. The crude residue was purified by flash chromatography (Hexane:EtOAc, 19:1) to afford dioxazoline **4** in 59% yield.

4 2,2,2-trifluoroethyl 2-(5-methyl-5-phenyl-1,4,2-dioxazol-3-yl)benzoate

Chemical Formula: $C_{18}H_{14}F_3NO_4$
Exact Mass: 365.09

Yield = 59% Colorless oil. R_f = 0.60 (4:1 Hexanes: Ethyl Acetate)

¹H NMR (500 MHz, Chloroform-*d*) δ 7.85 – 7.77 (m, 1H), 7.77 – 7.70 (m, 1H), 7.66 – 7.55 (m, 4H), 7.49 – 7.36 (m, 3H), 4.63 (dq, *J* = 12.6, 8.4 Hz, 1H), 4.49 (dq, *J* = 12.6, 8.4 Hz, 1H), 2.02 (s, 3H).

¹³C NMR (126 MHz, CDCl₃) δ 165.36, 157.63, 139.94, 132.19, 131.42, 130.27, 130.15, 129.84, 129.42, 128.63, 125.19, 123.14, 116.33, 61.34 (q, *J* = 36.9 Hz), 25.52.

IR(neat): 2975, 1746, 1494, 1292, 1163, 1123, 1097, 1010, 963, 698, 581 cm⁻¹

LRMS m/z (ESI APCI): calculated for $C_{18}H_{14}F_3NO_4$ [M+H] 366.1, found 366.1.

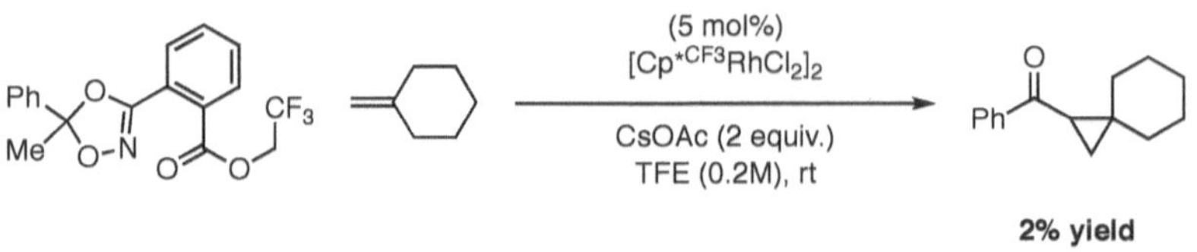

Figure 10.

Dioxazoline **4** (0.1 mmol), catalyst [Cp*^{CF3}RhCl$_2$]$_2$ (5 mol%, 0.005 mmol), and CsOAc (2 equiv., 0.2 mmol) were weighed in a 1-dram vial with a magnetic stirbar. TFE (0.2 M, 500 μL) was added followed by addition of alkene **2d** (1.2 equiv. 0.12 mmol). The vial was sealed with a screw-cap and stirred for 3 hours. Upon completion judged by TLC, the crude solution was diluted with EtOAc and partitioned with the addition of DI water. The aqueous layer was extracted three times with EtOAc and the combined organic extracts were filtered through a pad of celite® and Na$_2$SO$_4$ then concentrated. The yield of **3ad** was judged by the crude ^{1}H-NMR to be 2%.

Mechanistic hypothesis

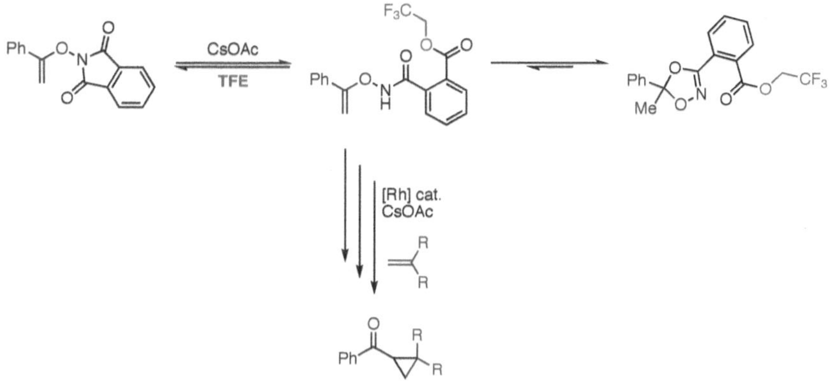

Figure 11.

Based on the experiment in figure 10, we conclude that the dioxazoline **4** does not

contribute significantly to the cyclopropanation reaction and is instead an off-cycle

intermediate.

A1.5 NMR Spectra

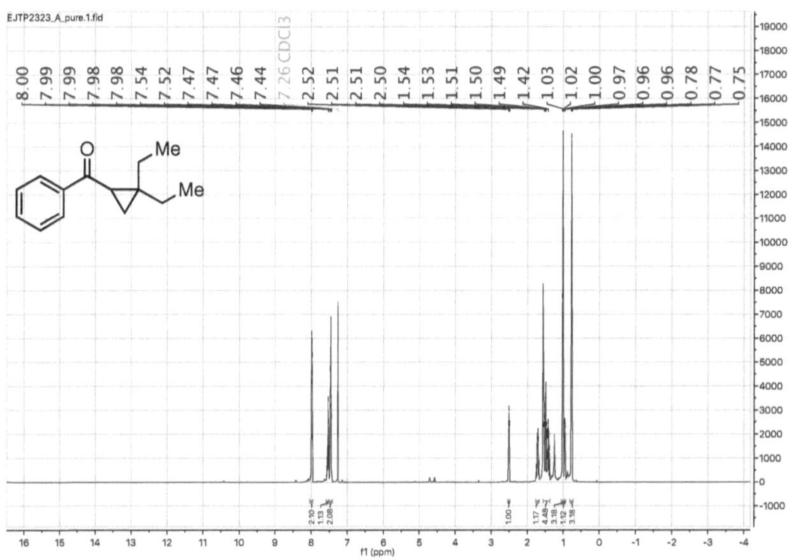

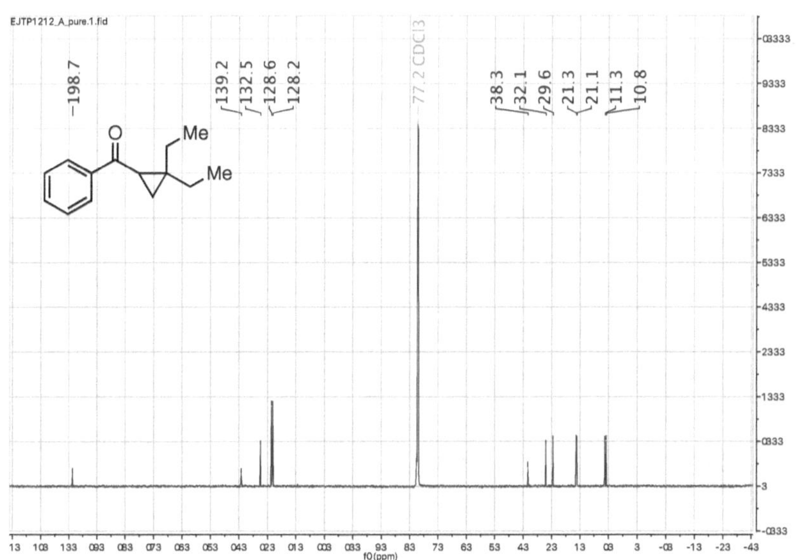

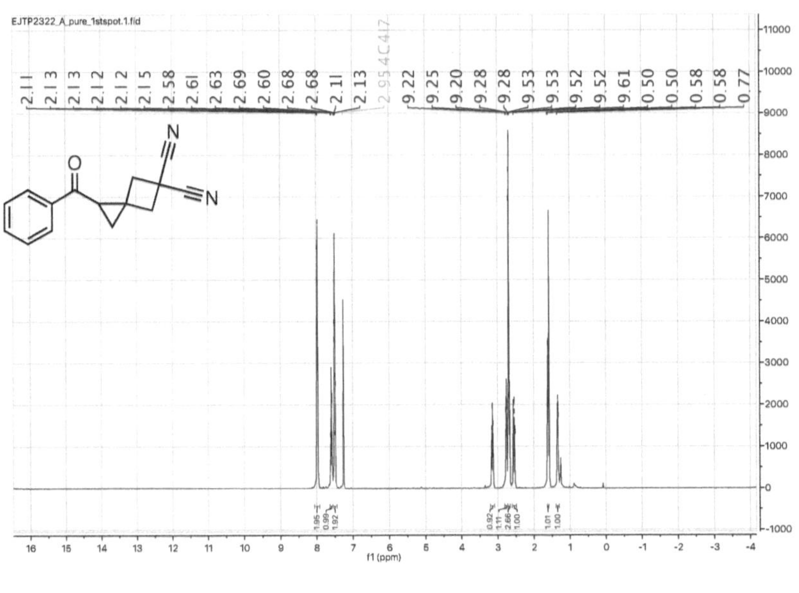

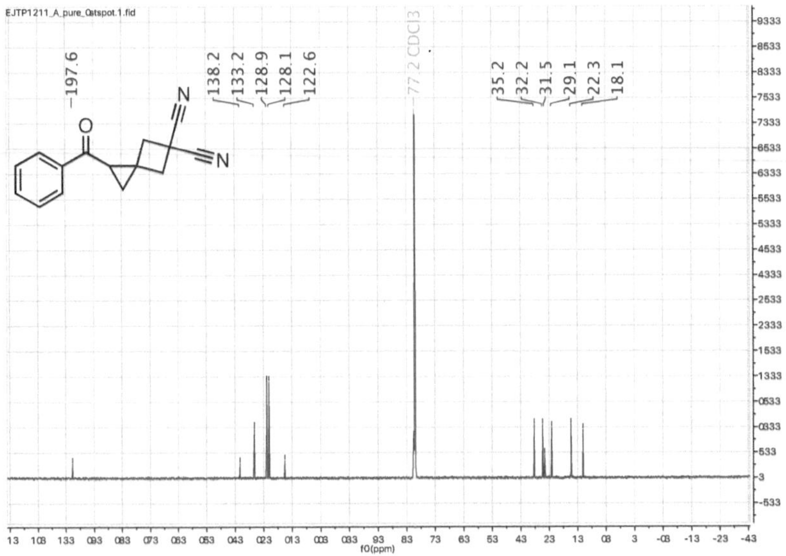

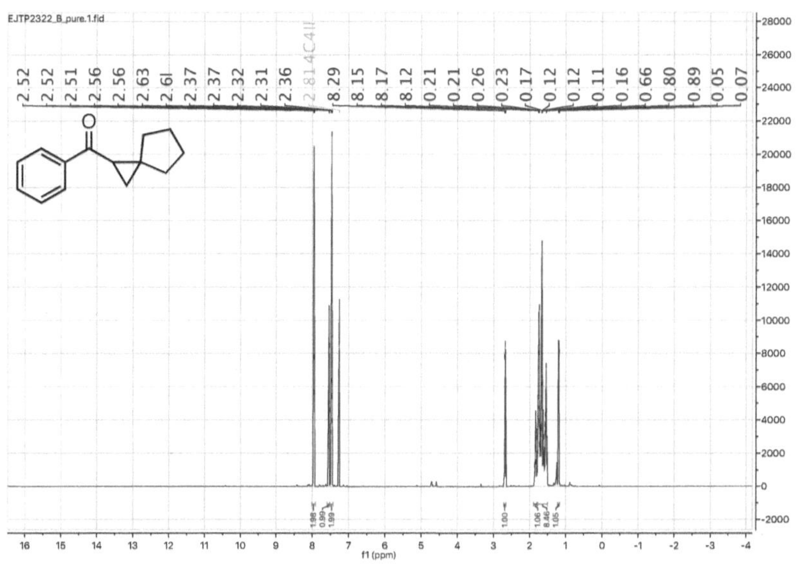

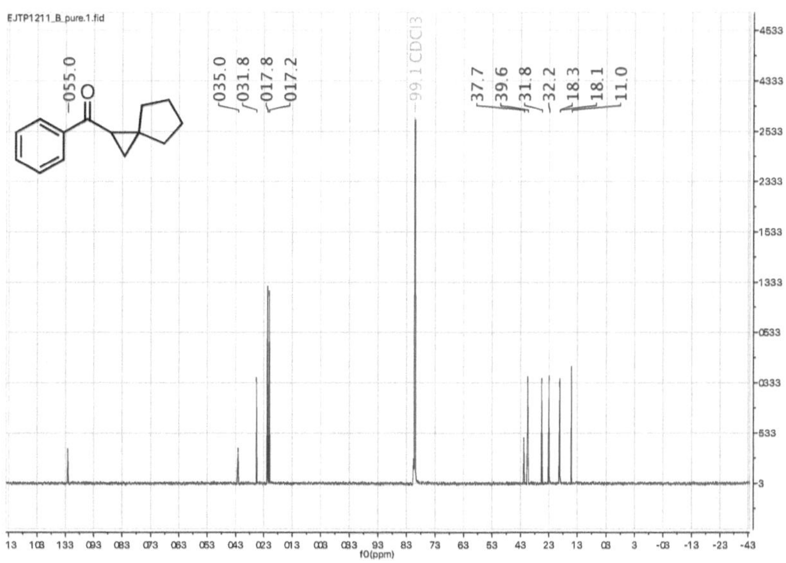

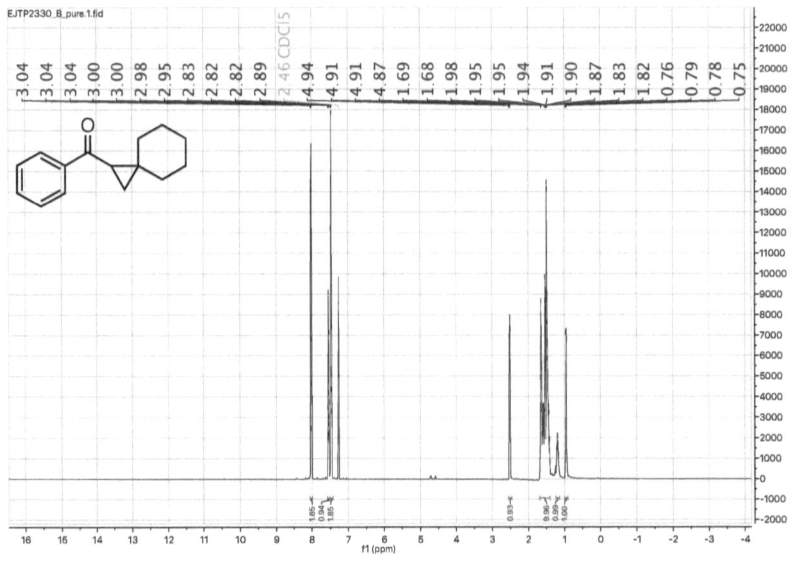

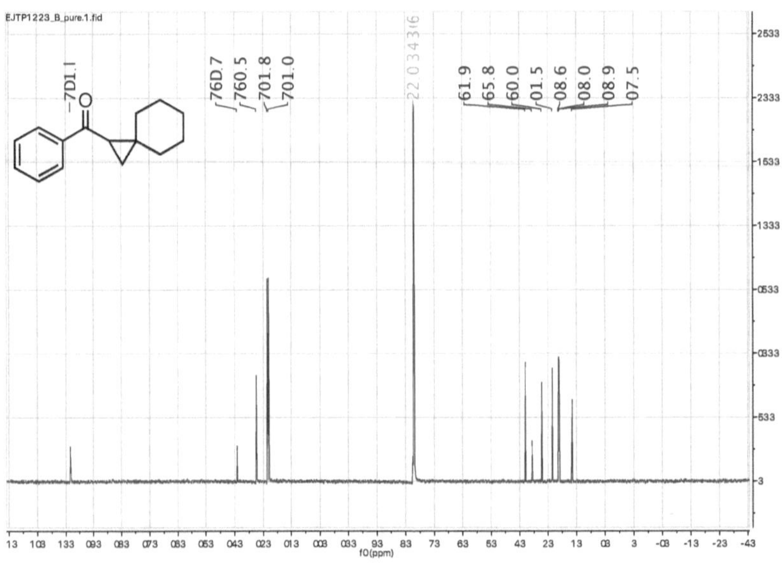

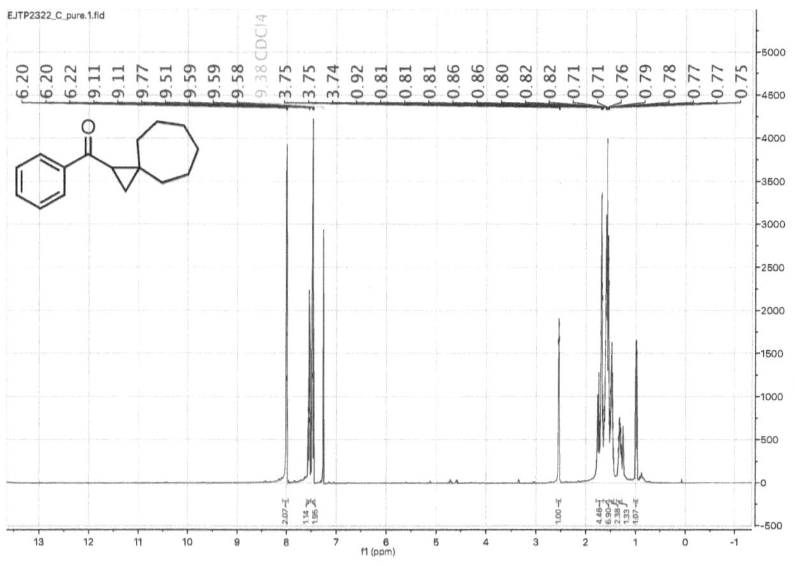

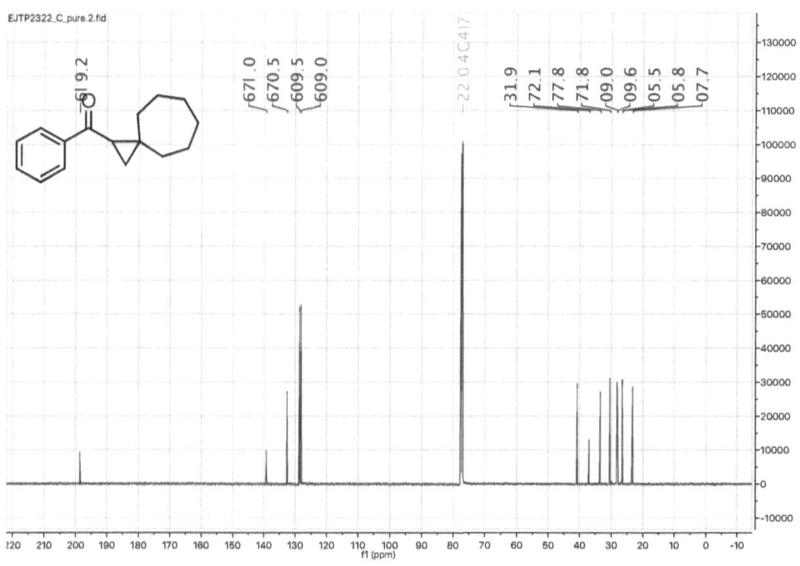

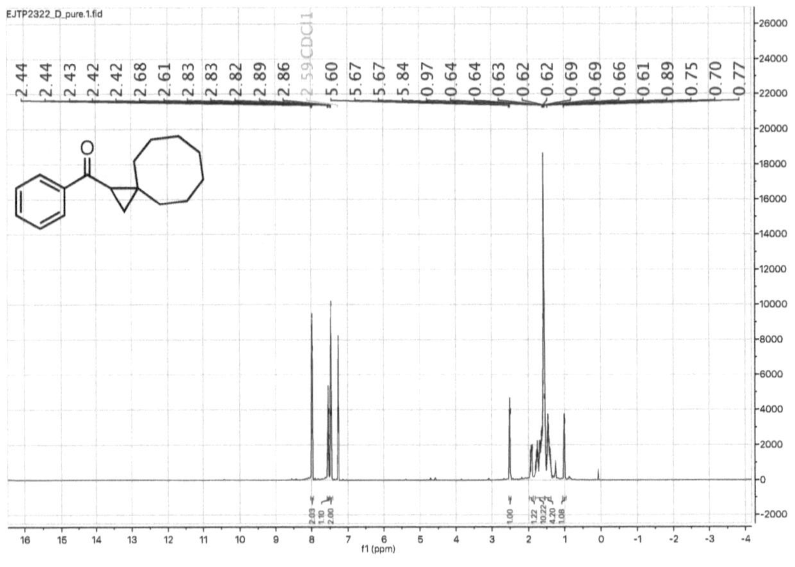

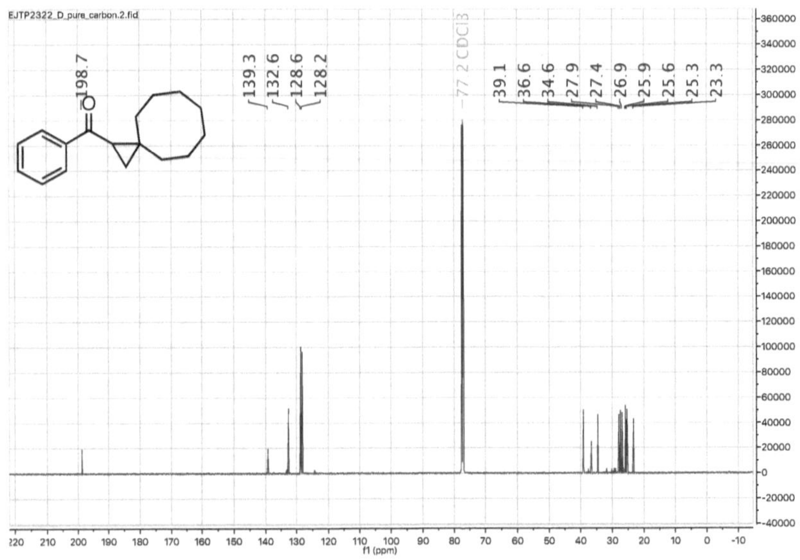

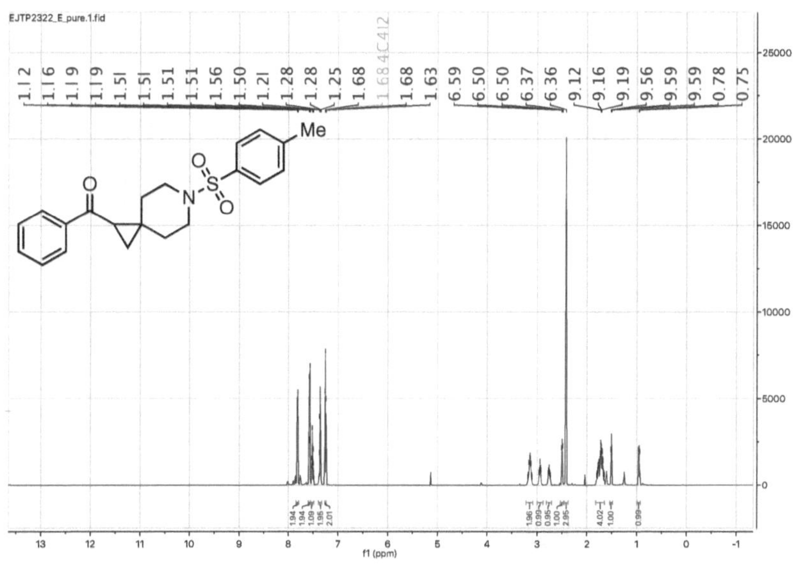

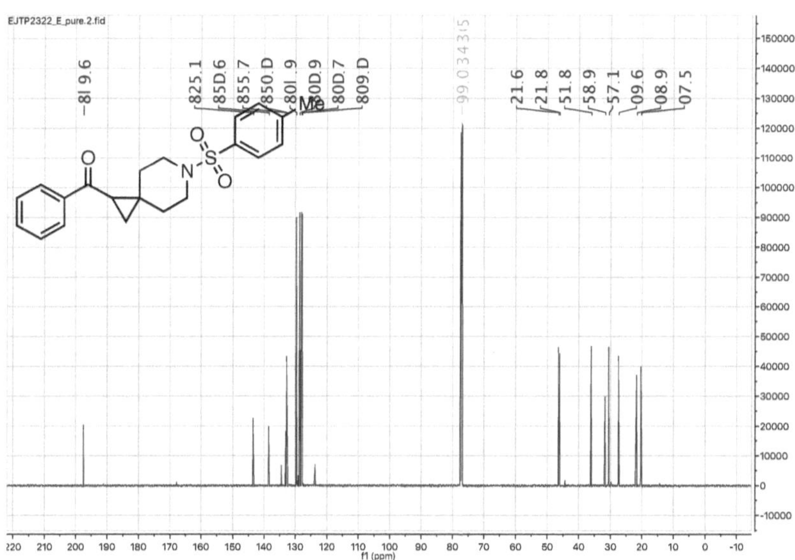

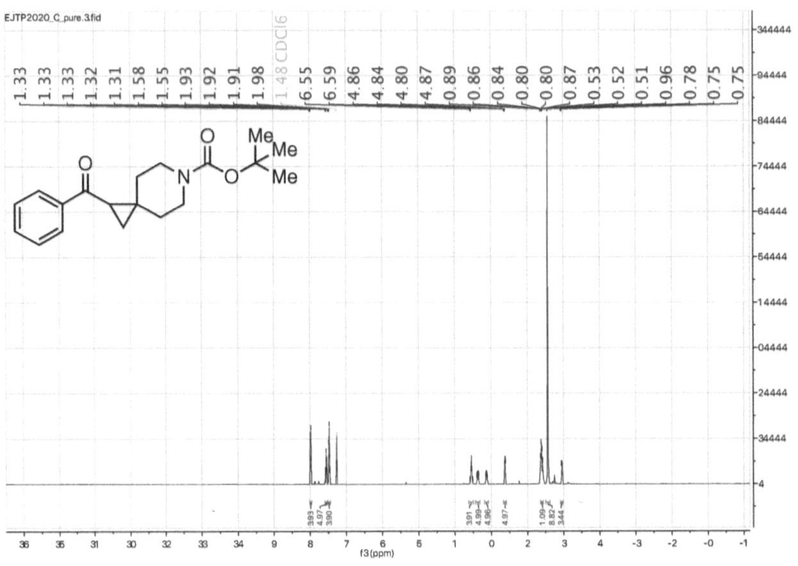

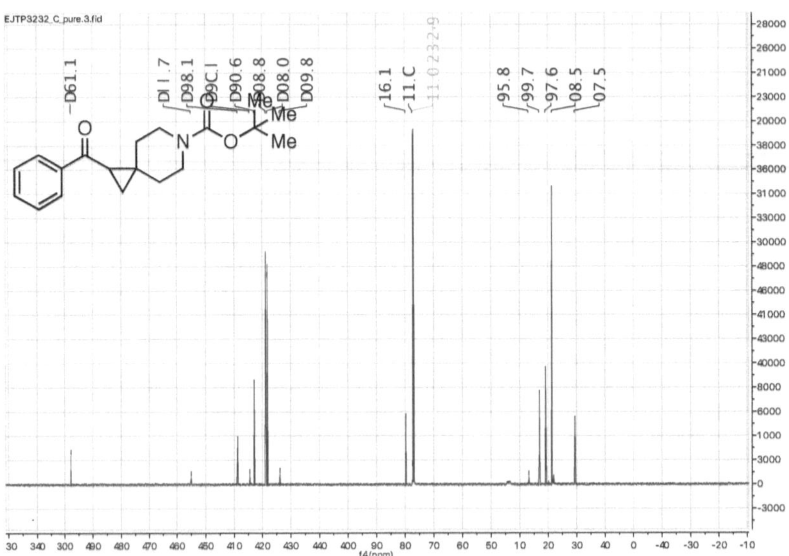

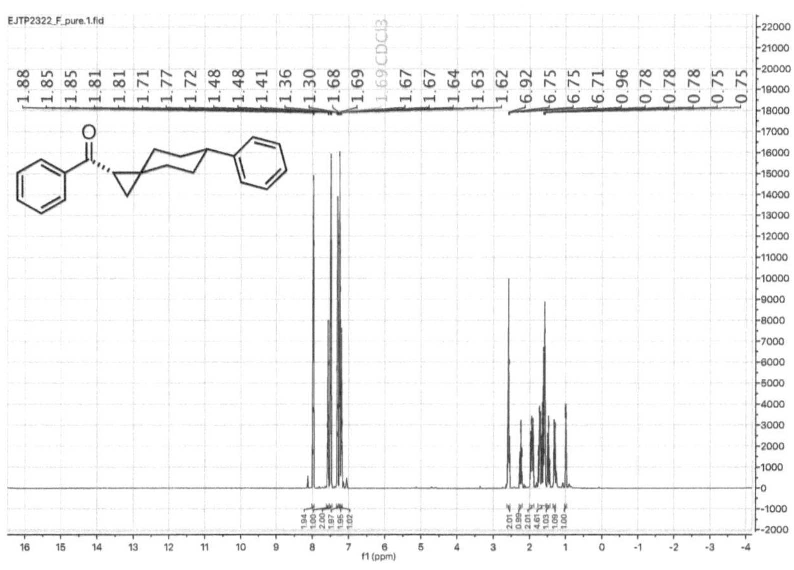

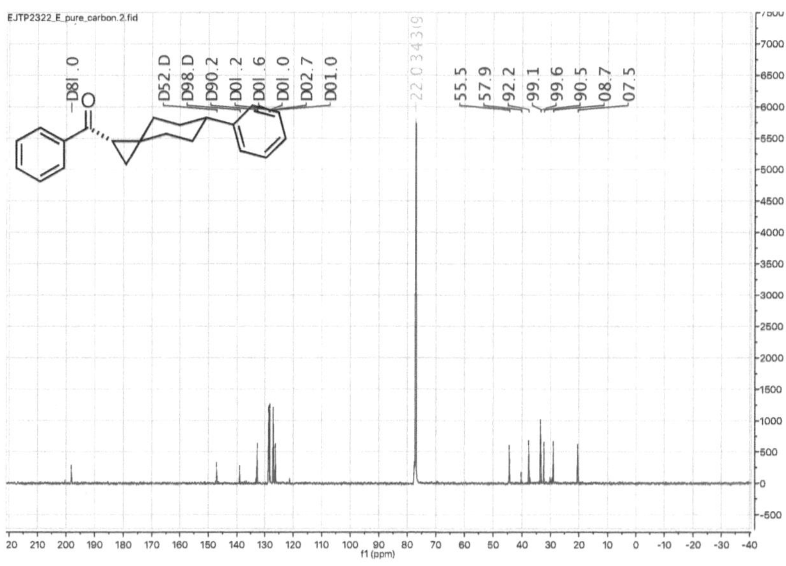

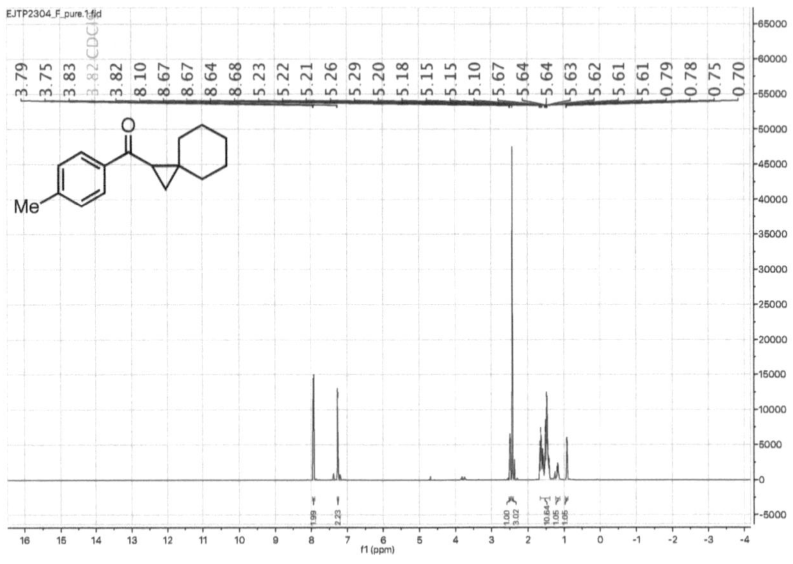

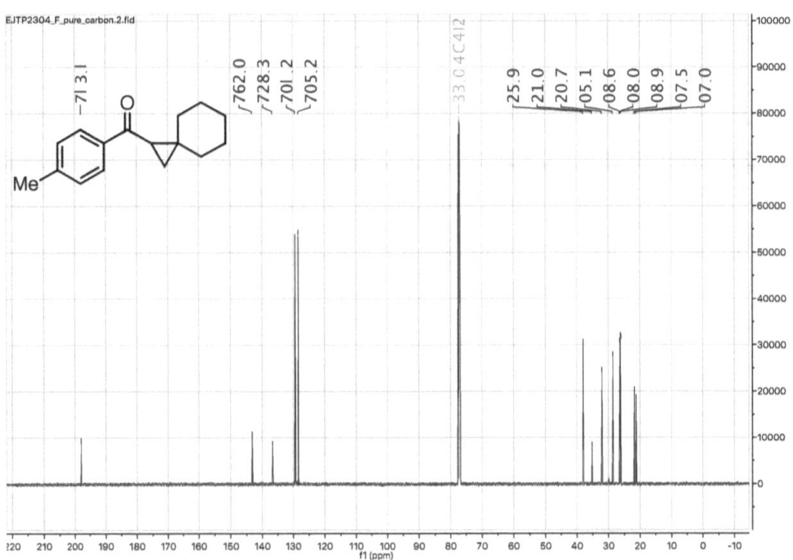

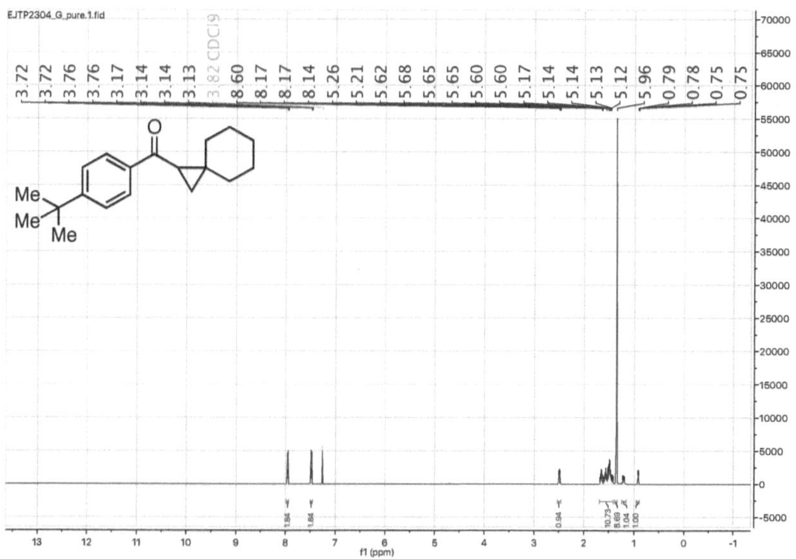

EJTP2304_G_pure.1.fid

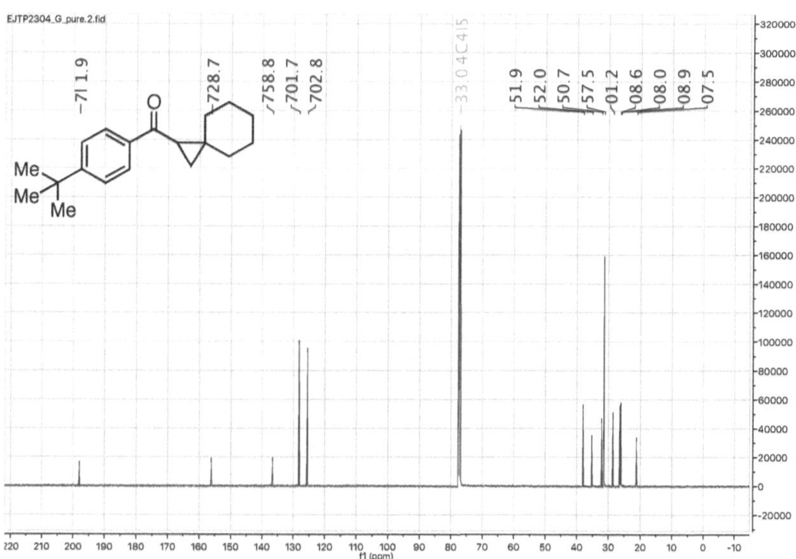

EJTP2304_G_pure.2.fid

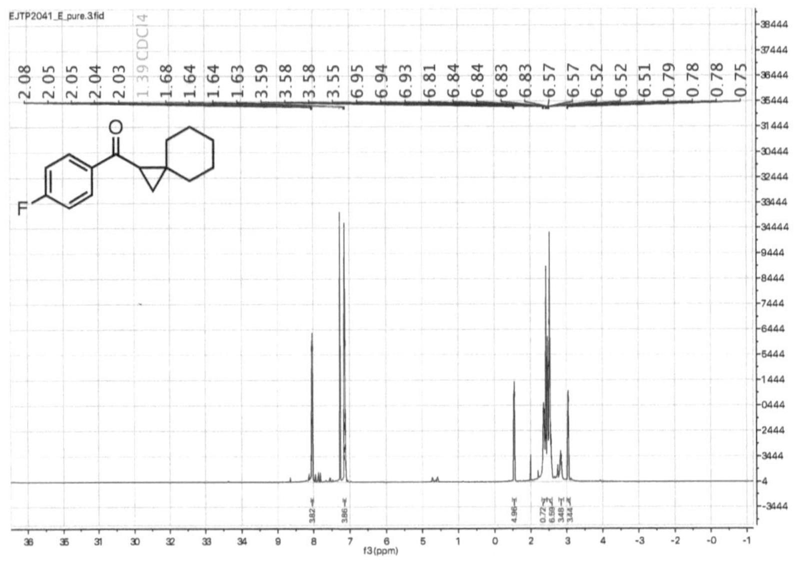

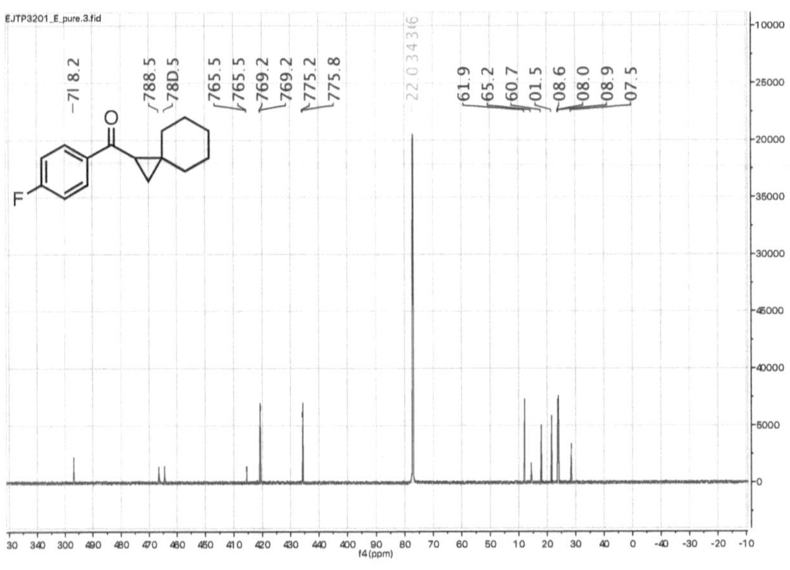

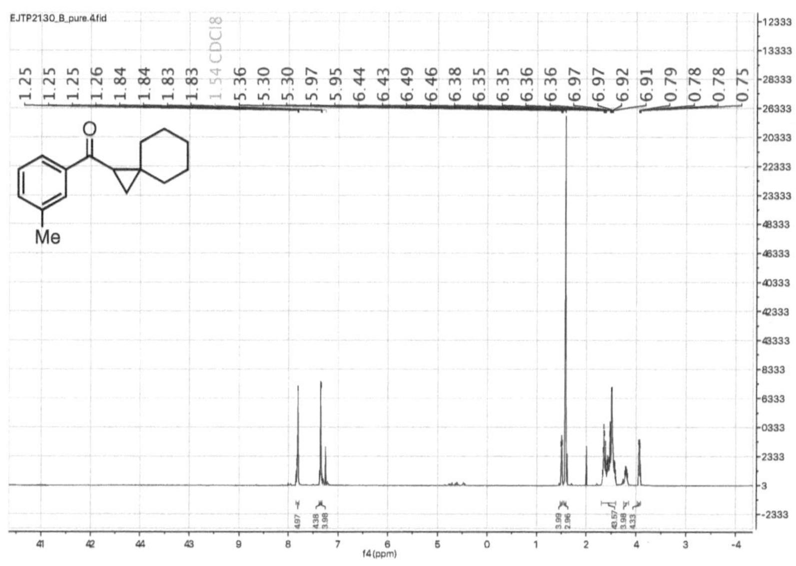

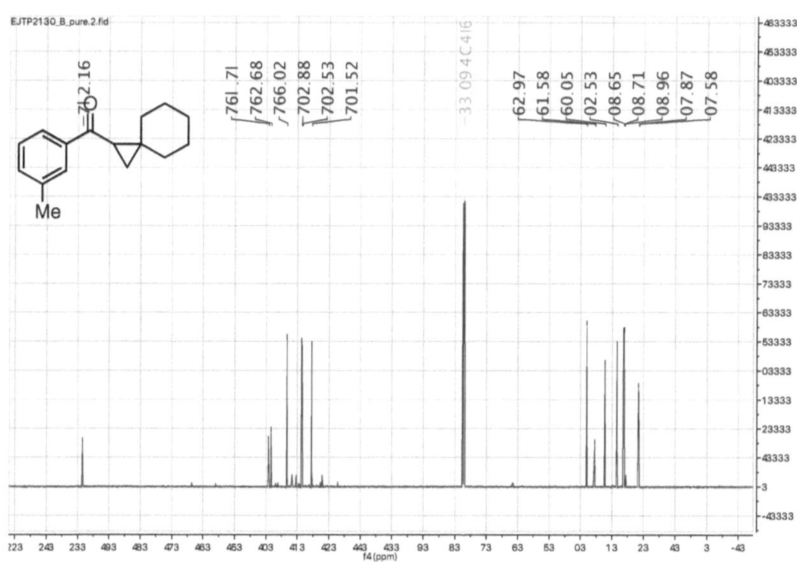

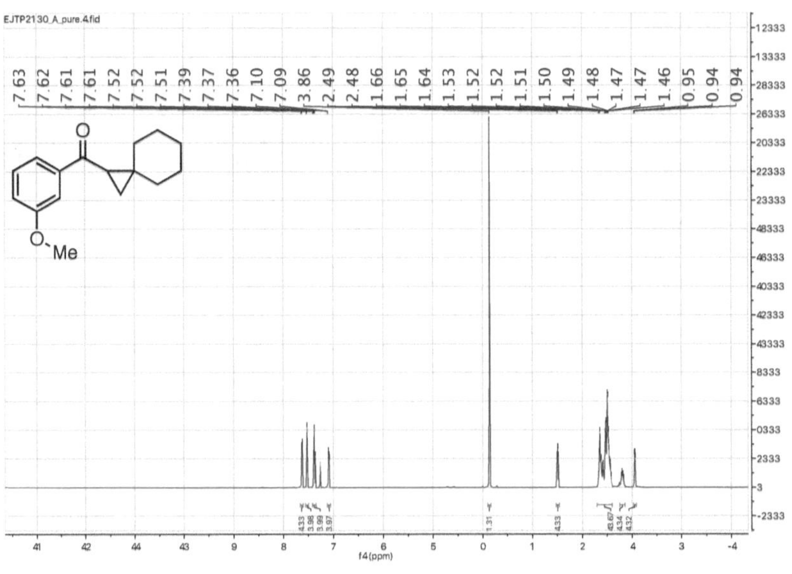

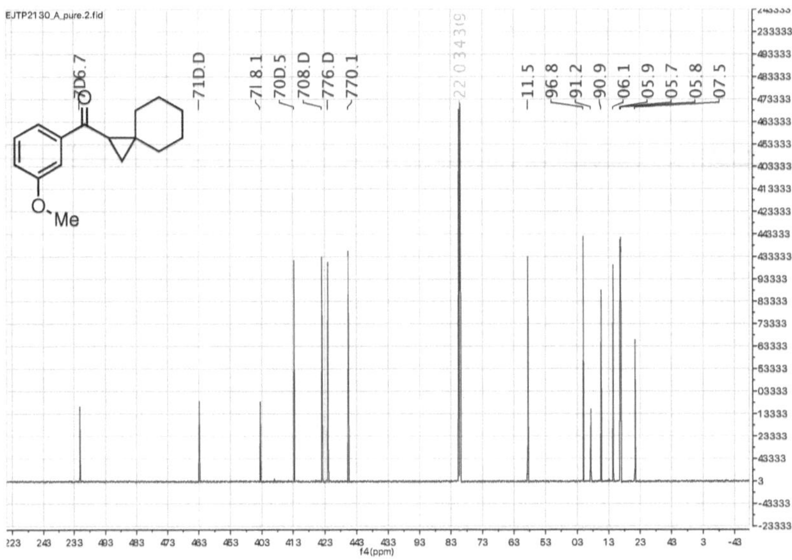

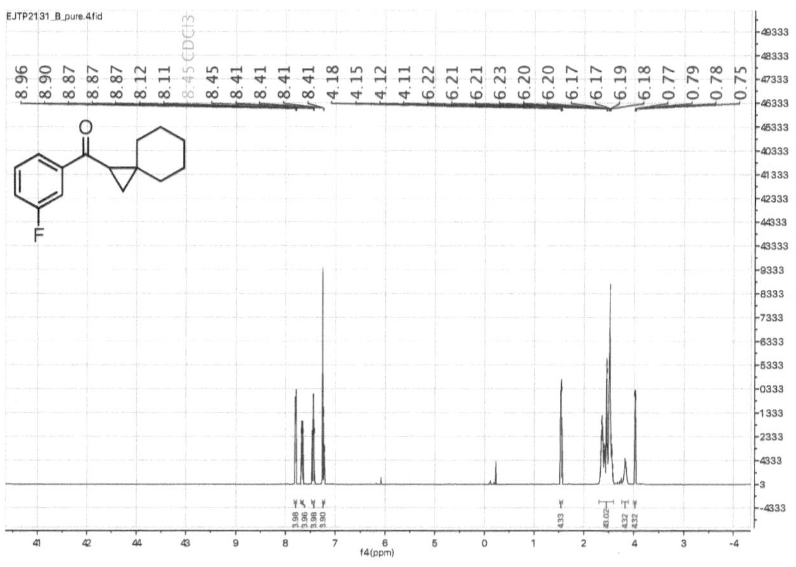

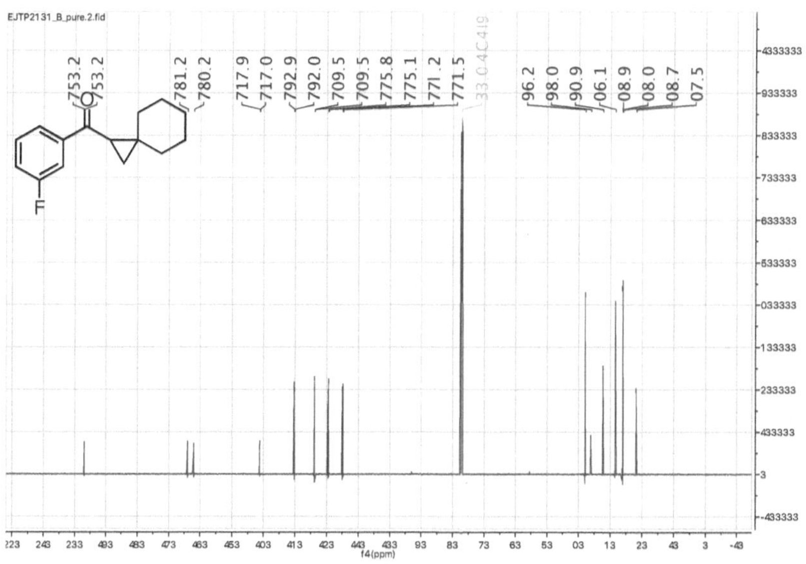

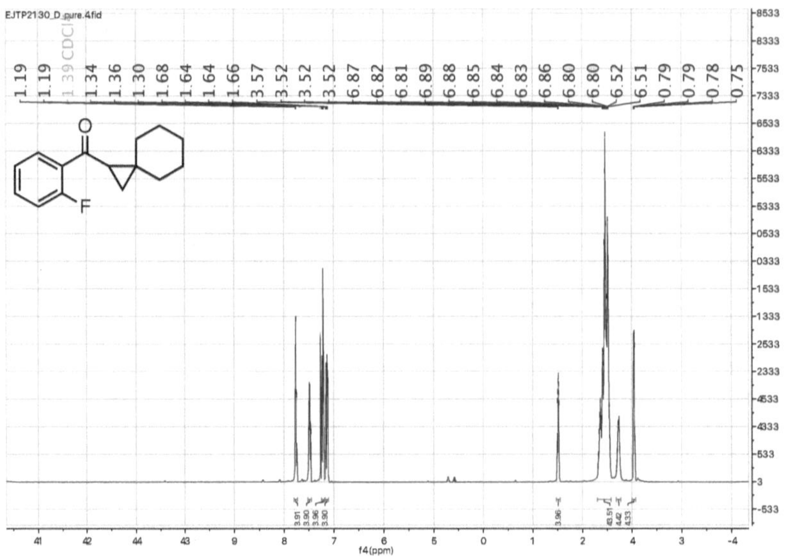

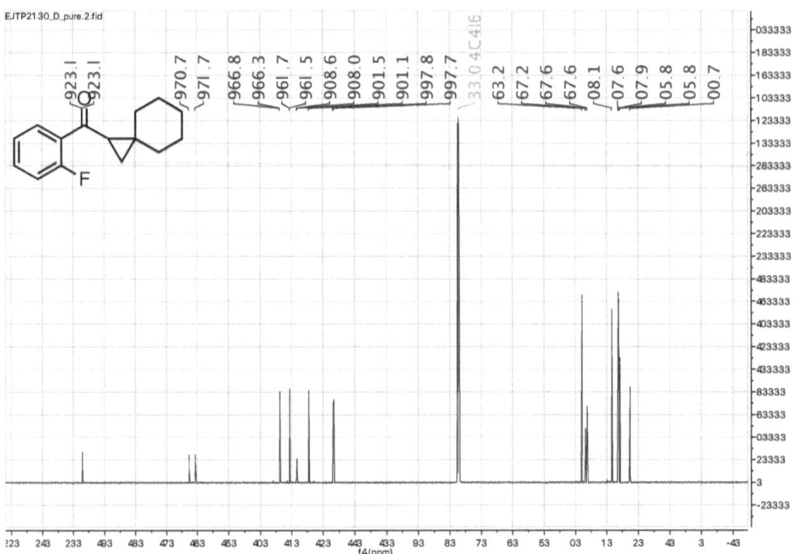

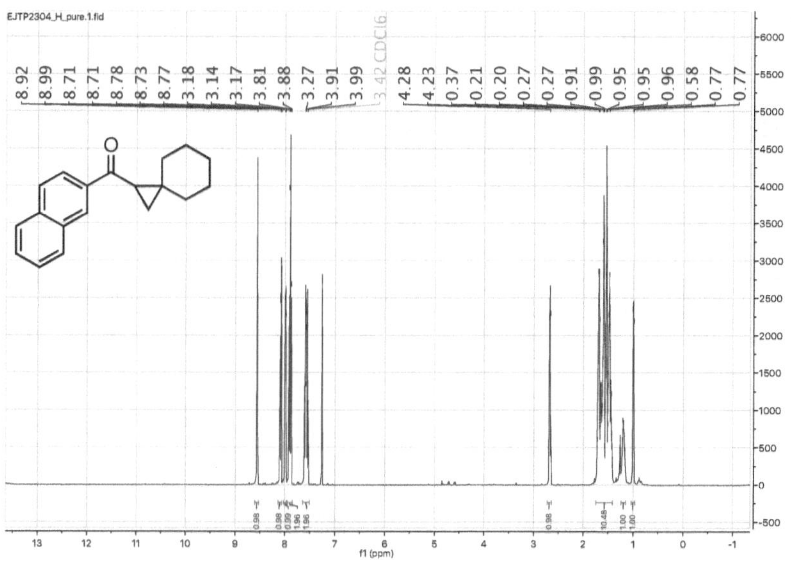

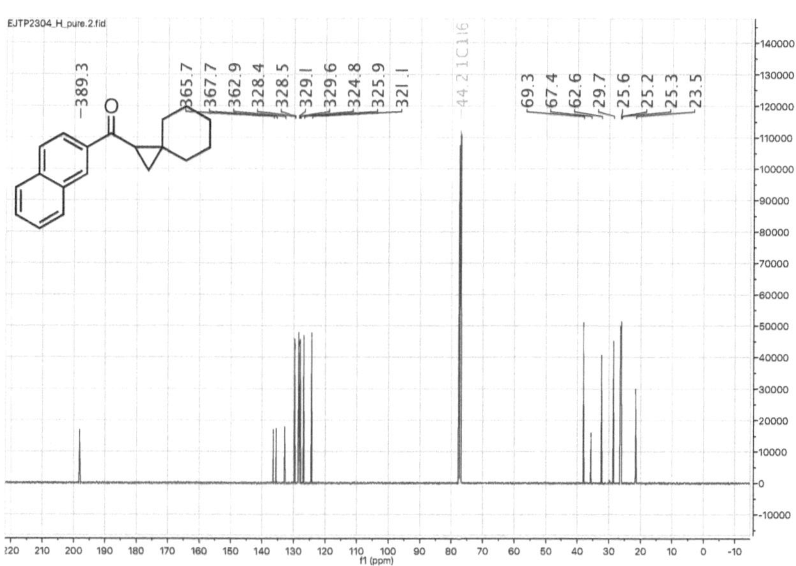

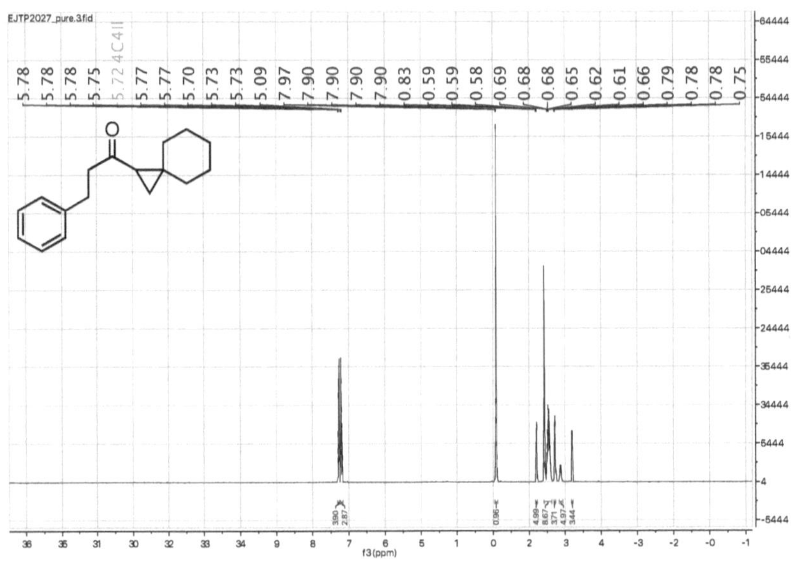

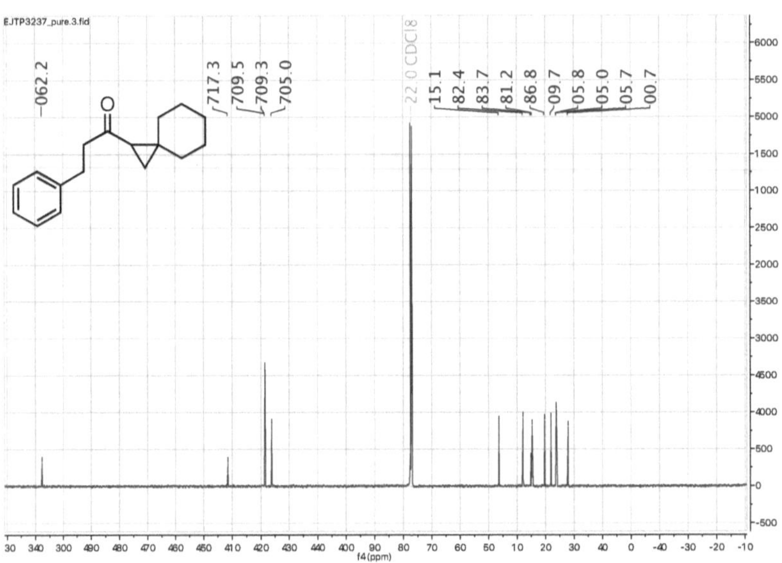

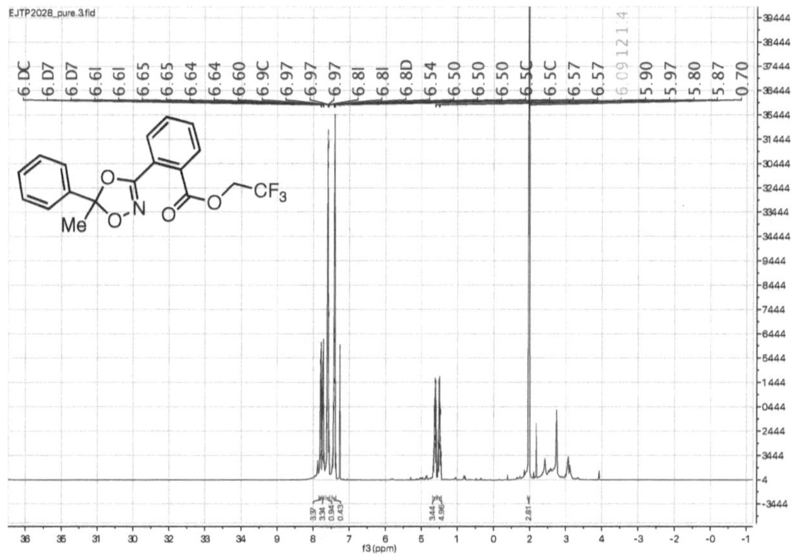

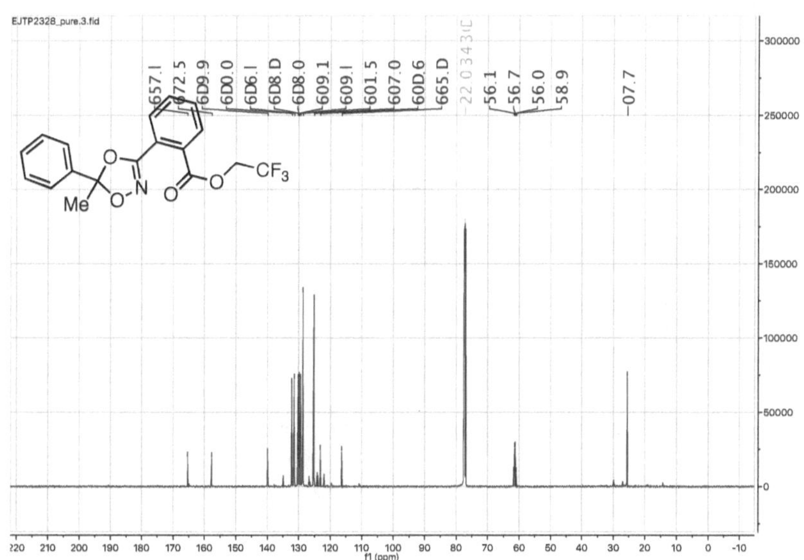

A1.6 References

---

(1) Still, W. C.; Kahn, M.; Mitra, A. *J. Org. Chem.* **1978**, *43*, 2923.

(2) Gassman, P. G.; Sowa, J. R. 1,2,3,4-Tetraalkyl-5-perfluoroalkyl-cyclopentadiene, di-(perfluoroalkyl)-trialkylcyclopentadiene and transition metal complexes thereof, U.S. Patent 5,245,064, Sep. 14, 1993.

(3) Phipps, E. J. T.; Rovis, T. *J. Am. Chem. Soc.*, **2019**, *141*, 6807.

(4) **1a-1i:** Piou, T.; Rovis, T. *J. Am. Chem. Soc.*, **2014**, *136*, 11292.

(5) **1j:** Duchemin, C.; Cramer, N. *Org. Chem. Front.*, **2019**, *6*, 209.

(6) **2e:** Kantorowski, E. J.; Borhan, B.; Nazarian, S.; Kurth M. J. *Tetrahedron Lett.* **1998**, *39*, 2483.

2f: Barluenga, J.; Fernández-Simón, J. L.; Cancellón, J. M.; Yus, M. *J. Chem. Soc. Perkin Trans.* **1988**, *1*, 3339.

2g and general procedure: Romanov-Michailidis, F.; Sedillo, K. F.; Neely, J. M.; Rovis, T. *J. Am. Chem. Soc.*, **2015**, *137*, 8892.

2h: Green, S. A.; Vásquez-Céspedes, S.; Shenvi, R. A. *J. Am. Chem. Soc.*, **2018**, *140*, 11317.

2i: Soulard, V.; Villa, G.; Vollmar, D. P.; Renaud, P. *J. Am. Chem. Soc.*, **2018**, *140*, 155.

– Appendix B –

Supporting Information for Chapter Three

Rh(III)-Catalyzed C–H Activation-Initiated Directed Cyclopropanation of Allylic Alcohols
Author: Erik J. T. Phipps, Tomislav Rovis
Publication: Journal of the American Chemical Society
Publisher: American Chemical Society
Date: May 1, 2019

PERMISSION/LICENSE IS GRANTED FOR YOUR ORDER AT NO CHARGE
This type of permission/license, instead of the standard Terms & Conditions, is sent to you because no fee is being charged for your order. Please note the following:

- Permission is granted for your request in both print and electronic formats, and translations.
- If figures and/or tables were requested, they may be adapted or used in part.
- Please print this page for your records and send a copy of it to your publisher/graduate school.
- Appropriate credit for the requested material should be given as follows: "Reprinted (adapted) with permission from (COMPLETE REFERENCE CITATION). Copyright (YEAR) American Chemical Society." Insert appropriate information in place of the capitalized words.
- One-time permission is granted only for the use specified in your request. No additional uses are granted (such as derivative works or other editions). For any other uses, please submit a new request.

If credit is given to another source for the material you requested, permission must be obtained from that source.

Rh(III)-Catalyzed C–H Activation-Initiated Diastereoselective Directed Cyclopropanation of Allylic Alcohols

Supporting Information

Erik J.T. Phipps and Tomislav Rovis*

Table of Contents

A2.1 General Methods

All reactions were carried out in oven-dried glassware with magnetic stirring. ACS grade TFE and reagents were purchased from TCI, Strem, Alfa Aesar, and Sigma-Aldrich and were used without further purification. Dichloromethane, tetrahydrofuran, diethyl ether were degassed with argon and passed through two columns of neutral alumina. Column chromatography was performed on SiliCycle® SilicaFlash® P60, 40-63 μm 60 Å and in general were run using flash techniques.[1] Thin layer chromatography was performed on SiliCycle® 250 μm 60 Å plates. Visualization was accomplished with UV light (254 nm). ^{1}H, ^{19}F, and ^{13}C NMR spectra were collected at ambient temperature in CDCl$_3$ on Bruker 300Hz, 400 MHz, or 500MHz spectrometers. Chemical shifts are expressed as parts per million (δ, ppm) and are referenced to the residual solvent peak of chloroform([1]H = 7.26 ppm; [13]C = 77.2 ppm). Scalar coupling constants (*J*) are quoted in Hz. Multiplicity is reported as follows: s = singlet, d = doublet, t = triplet, q = quartet, m= multiplet). Mass spectra were obtained on a Waters (LRMS). Infrared (IR) spectra were obtained with neat samples on a Bruker Tensor 27 FT-IR spectrometer with OPUS software. Typically, the experiment consisted measuring the transmission in 16 scans in the region from 4000 to 400 cm[-1].

A2.2 General Procedure for Starting Materials

A. *Synthesis of [Cp*CF3RhCl2]2 Catalyst[2]*

Synthesis of 1,2,3,4-tetramethyl-5-(trifluoromethyl)cyclopenta-1,3-diene (+ isomers)

Following a reported procedure, Li wire (1.291 g, 186 mmol, 4 equiv.) was cut into ~5

mm size pieces and added to Et_2O (2.85 M, 69 mL) in a 250-mL 3-neck flask with a

magnetic stir bar and cooled to 0 °C in an ice bath. 2-bromo-2-butene (cis + trans)

(9.7 mL, 95.3 mmol, 2.05 equiv.) diluted with 10 mL Et_2O was added dropwise over 10

minutes. The heterogeneous mixture was stirred for 2 hours then cooled to -40 °C

(MeCN, Dry Ice bath). Ethyl trifluoroacetate (5.4 mL, 46.5 mmol, 1 equiv.) diluted

with 5 mL Et_2O was added dropwise over 10 minutes. The solution was stirred for an

additional 90 minutes. The solution was quenched with 20 mL of 2 M HCl solution

and diluted with 100 mL DI H_2O. The solution was transferred to a separatory funnel

and the layers separated. The aqueous layer was extracted three times with Et_2O. The

organic layers were combined and washed with saturated sodium bicarbonate, water,

and brine then dried over Na_2SO_4 and concentrated. The resulting yellow liquid was vacuum distilled to give the intermediate alcohol, a clear liquid, in 43% yield (4.1734 g).

The intermediate alcohol (1.0 g, 4.8 mmol, 1 equiv.) was dissolved in DCM (0.16 M, 30 mL) in a 50-mL flask equipped with a magnetic stir bar and cooled to 0 °C in an ice bath. Methanesulfonic acid (3.1 mL, 48 mmol, 10 equiv.) was quickly added and the solution was stirred for 5 minutes. The resulting dark red solution was then poured into 50 mL of cooled DI H_2O. The solution was transferred to separatory funnel and the layers were separated. The aqueous layer was extracted three times with DCM. The organic layers were combined and washed with saturated sodium bicarbonate, dried over dried over Na_2SO_4 and concentrated. HCp*[CF3] (+ isomers) was purified by flash chromatography (Hexanes) and afforded in 69% yield (1.3374 g)

Synthesis of [Cp[CF3]RhCl$_2$]$_2$*

$RhCl_3$ • 3 H_2O + (CF$_3$, Me, Me, Me, Me cyclopentadiene) + isomers → (MeOH, Δ) → [Cp*[CF3]RhCl$_2$]$_2$

From a reported procedure, in a 250-mL flask equipped with a magnetic stir bar and a reflux condenser under N_2 atmosphere was added $RhCl_3$ • 3 H_2O (700 mg, 2.6 mmol, 1 equiv.), MeOH (140 mL, 0.019 M), and HCp*[CF3] (1.3374 g, 7.28 mmol, 2.7 equiv.).

The solution was refluxed under N_2 atmosphere for 3 days where a dark red precipitate was visible on the sides of the flask. The reaction was cooled to 0 °C in an ice bath and the precipitate was filtered and washed with EtOH two additional times. The resulting red solid was collected and dried to afford 72% yield (1.33 g). The

B. Synthesis of N-enoxyphthalimide Substrates

Method A[3]

Synthesis of (1,2-dibromoethyl)arenes

Styrene (1 equiv.) in DCM (0.5M) was cooled to 0 °C and Br_2 (1.2 equiv.) was added via syringe and stirred at 0 °C for ~1 hour. The solution was quenched with sat. $Na_2S_2O_3$ until the solution became colorless. The resulting solutions was then filtered through a pad a celite® and washed with DCM. The layers were then separated and the aqueous layer was extracted with DCM. The combined organic layers were then washed with brine, dried over Na_2SO_4, and concentrated. The resulting white solid was directly carried on without purification.

Synthesis of α-bromostyrenes

(1,2-dibromoethyl)arenes (1 equiv.) was stirred in a 0.25M solution of 1:1 methanol

and THF at room temperature. Potassium carbonate (2 equiv.) was added and the

solution stirred until the reaction was judged complete by TLC (~3 hrs.). The reaction

was then quenched with D.I. water and the volatiles were removed. The resulting

aqueous layer was extracted with ether and the combined organic layers were then

washed with brine, dried over Na_2SO_4, and concentrated.

The resulting oil was directly carried on without purification.

Synthesis of (1-arylvinyl) boronic acids

α-bromostyrene in dry diethyl ether was put under inert atmosphere in a 2-neck flask

and cooled to -78 °C. A 1.7M solution of t-BuLi in pentanes (2.1 equiv.) was added

dropwise and the solution was stirred at -78 °C for 30 minutes. Tri-isopropylborate

(1.2 equiv.) was added dropwise to the solution over 30 minutes. After the addition

was complete, the solution was stirred at -78 °C for 2 hours after which the solution was removed from the cold bath and stirred at room temperature overnight. To the resulting yellow-orange solution was added 1M HCl solution and was stirred for 2 hours. The layers were separated and the aqueous layer was extracted with ether. The combined organic layers were then washed with 1M NaOH solution and the layers were separated. The <u>aqueous</u> layer was acidified to pH≈1 and extracted with ethyl acetate. The combined organic layers were then washed with brine, dried over Na_2SO_4, and concentrated. The crude product was directly carried on without purification.

Synthesis of N-enoxyphthalimides

Boronic acid (2 equiv.), copper(II) acetate (1 equiv.), *N*-hydroxyphthalimide (1 equiv.), and anhydrous sodium sulfate (4 equiv.) were combined in a flask and diluted with 1,2-dichloroethane to form a 0.1M solution of *N*-hydroxyphthalimide. Pyridine (3 equiv.) was added via syringe and the solution was stirred at room temperature open to air for 2 days. At the end of the stirring period, the volatiles were removed and the resulting solids were purified by column chromatography. The purified solids were then used in the cyclopropanation reactions.

Method B[4]

$$R\text{---}\!\!\!\equiv \quad + \quad HO\text{-}NPhth \quad \xrightarrow[\text{DCE, 90 °C}]{\text{[PPh}_3\text{AuTFA] (5 mol\%)}} \quad R\text{-C(=O)-O-NPhth}$$

Following a reported procedure, alkyne (3 equiv.), *N*-hydroxphthalimide (1 equiv.), and

Au catalyst (5 mol%) were combined in a 1.5 dram vial in the glove box under Ar and

dissolved in 1,2-DCE (0.2M). The vial was sealed and removed from the glovebox and

placed in an aluminum heating block overnight at 90 °C. The reaction was then cooled

to room temperature, diluted with DCM and passed through a pad of Celite®. The

solvent was removed and the crude residue was purified by column chromatography

(19:1, Hex:EtOAc).

Compounds Synthesized by Method A

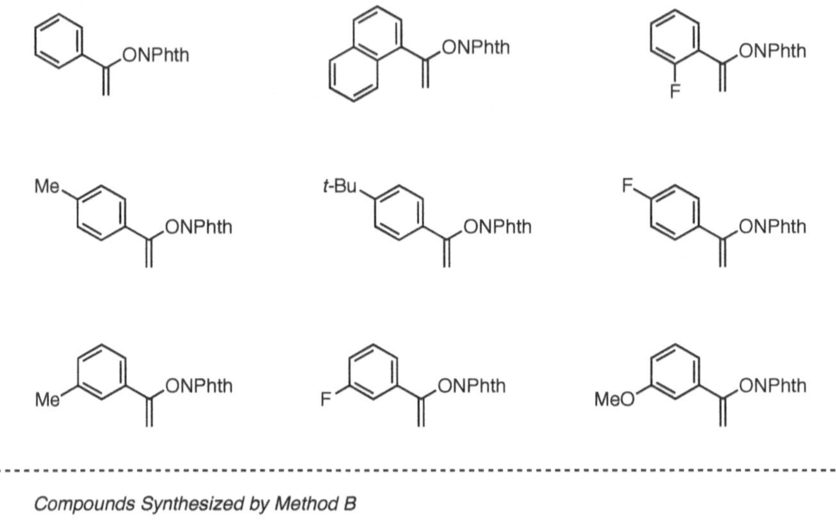

Compounds Synthesized by Method B

The known compounds are consistent with the literature precedents.

Allylic alcohols were purchased from commercial suppliers unless noted below:

This procedure was performed according to literature precedent.[5, 6, 7]

This procedure was performed similar to literature precedent. The crude mixture was

purified by flash chromatography (9:1→4:1, Hex:EtOAc). The compound was

consistent with the literature.[8]

This procedure was performed according to literature precedent and the compound was consistent with the literature.[9]

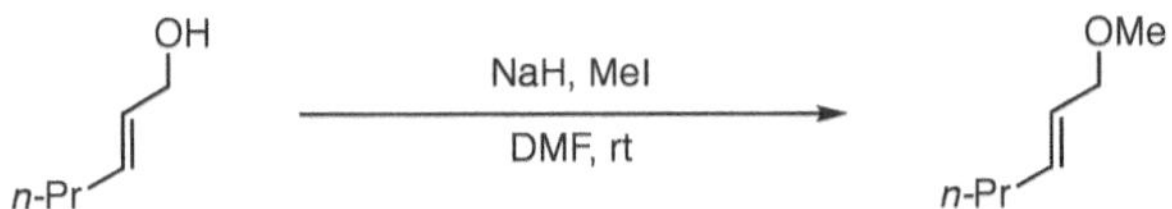

This procedure was performed according to literature precedent and the compound was consistent with the literature.[10]

This procedure was performed according to the literature precedent and the compounds were consistent with the literature.[11]

A2.3 General Procedure for the Cyclopropanation Reaction and Characterization of Products

N-enoxyphthalimide (0.12 mmol), catalyst [Cp*^{CF3}RhCl$_2$]$_2$ (5 mol%, 0.006 mmol, 4.4 mg), and KOPiv (2 equiv., 0.24 mmol, 33.6 mg) were weighed in a 1-dram vial with a magnetic stirbar. Cooled TFE (0.2 M, 600 μL) was added followed by allylic alcohol (1.2 equiv., 0.144 mmol). The vial was sealed with a screw-cap and placed in an aluminium block cooled to 0 °C surrounded by ice in an insulated box and stirred for 16 hours. Upon completion judged by TLC, TFE was removed by rotary evaporation and the residue was taken up in EtOAc and filtered through a silica plug flushing with EtOAc. The filtrate was concentrated to ~1mL and transferred to a 1.5-dram vial where the solution was partitioned with the addition of 10% NaOH solution. The aqueous layer was extracted three times with EtOAc and the combined organic extracts were filtered through a pad of celite® and Na$_2$SO$_4$ then concentrated. The crude residue was purified by flash chromatography (Hexane:EtOAc, 19:1→9:1→4:1) to afford the cyclopropane product.

3aa *2-(hydroxymethyl)-3-propylcyclopropyl)(phenyl)methanone*

Chemical Formula: $C_{14}H_{18}O_2$

Y = 81%. Yellow oil. R_f = 0.22 (4:1 Hexanes:EtOAc)

^{1}H NMR (500 MHz, Chloroform-*d*) δ 8.05 – 7.94 (m, 2H), 7.56 (t, *J* = 7.4 Hz, 1H), 7.47 (t, *J* = 7.6 Hz, 2H), 3.95 (dd, *J* = 12.0, 4.7 Hz, 1H), 3.76 (dd, *J* = 12.0, 8.3 Hz, 1H), 2.55 (dd, *J* = 8.4, 5.0 Hz, 1H), 2.14 (s, 1H), 1.89 – 1.81 (m, 1H), 1.78 – 1.71 (m, 1H), 1.45 (tdd, *J* = 13.4, 10.7, 4.8 Hz, 4H), 0.93 (t, *J* = 6.9 Hz, 3H).

^{13}C NMR (126 MHz, CDCl$_3$) δ 200.2, 138.6, 133.0, 128.7, 128.2, 60.0, 35.6, 35.4, 30.3, 28.5, 22.4, 14.0.

IR(neat) 3456, 2924, 1660, 1453, 1228, 1019, 700 cm^{-1}

LRMS m/z (ESI APCI) calculated for $C_{14}H_{18}O_2$ [M+H] 219.1, found 219.1.

3ab *2-(hydroxymethyl)-3-propylcyclopropyl) (phenyl)methanone*

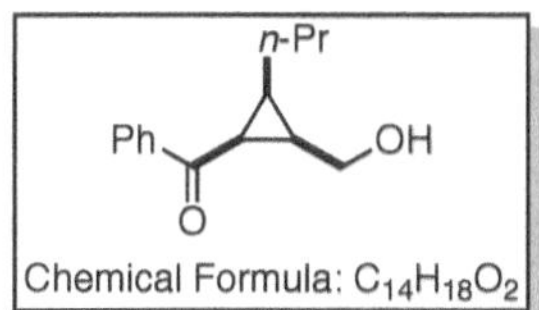

Y = 62%. Pale-yellow oil. R_f = 0.22 (4:1 Hexanes:EtOAc)

¹H NMR (500 MHz, Chloroform-*d*) δ 8.03 – 7.95 (m, 2H), 7.61 – 7.54 (m, 1H), 7.46 (dd, *J* = 8.4, 7.0 Hz, 2H), 4.06 (dd, *J* = 7.9, 2.7 Hz, 2H), 2.73 (dd, *J* = 9.4, 7.9 Hz, 1H), 2.45 (s, 1H), 1.90 – 1.72 (m, 2H), 1.55 (ddt, *J* = 13.9, 8.4, 6.9 Hz, 1H), 1.49 – 1.40 (m, 1H), 1.35 – 1.25 (m, 2H), 0.85 (t, *J* = 7.4 Hz, 3H).

¹³C NMR (126 MHz, CDCl₃) δ 200.7, 138.9, 133.1, 128.7, 128.3, 58.7, 28.0, 28.0, 26.5, 25.5, 23.1, 14.0.

IR(neat) 3433, 2957, 1680, 1449, 1209, 1020, 699 cm⁻¹

LRMS m/z (ESI APCI) calculated for $C_{14}H_{18}O_2$ [M+H] 219.1, found 219.1.

3ba *2-(hydroxymethyl)-3-propylcyclopropyl) (p-tolyl)methanone*

Me
n-Pr
OH
O
Chemical Formula: C₁₅H₂₀O₂

Y = 72%. Yellow oil. R_f = 0.22 (4:1 Hexanes:EtOAc)

¹H NMR (500 MHz, Chloroform-*d*) δ 7.90 (d, *J* = 8.2 Hz, 2H), 7.27 (d, *J* = 7.5 Hz, 2H), 3.94 (dd, *J* = 12.3, 4.8 Hz, 1H), 3.80 – 3.71 (dd, *J* = 9.0, 5.8 Hz, 1H), 2.52 (dd, *J* = 8.4, 5.1 Hz, 1H), 2.42 (s, 3H), 2.19 (s, 1H), 1.86 – 1.79 (m, 1H), 1.72 (tdd, *J* = 8.3, 6.5, 4.6 Hz, 1H), 1.52 – 1.37 (m, 4H), 0.98 – 0.89 (t, *J* = 7.0 Hz, 3H).

¹³C NMR (126 MHz, CDCl₃) δ 199.8, 143.8, 136.1, 129.4, 128.4, 60.1, 35.5, 35.4, 30.1, 28.3, 22.4, 21.8, 14.0.

IR(neat) 3441, 2957, 2923, 1663, 1607, 1454, 1233, 1179, 103, 665 cm⁻¹

LRMS m/z (ESI APCI) calculated for $C_{15}H_{20}O_2$ [M+H] 233.2, found 233.2.

3ca *(4-(tert-butyl)phenyl)-2-(hydroxymethyl)-3-propylcyclopropyl)methanone*

t-Bu, n-Pr, OH, O

Chemical Formula: $C_{18}H_{26}O_2$

Y = 76%. Pale-yellow oil. R_f = 0.24 (4:1, Hexanes:EtOAc).

^{1}H NMR (500 MHz, Chloroform-*d*) δ 7.95 (d, *J* = 8.5 Hz, 1H), 7.49 (d, *J* = 8.5 Hz, 2H), 3.98 – 3.91 (m, 1H), 3.81 – 3.72 (m, 1H), 2.54 (dd, *J* = 8.4, 5.0 Hz, 1H), 2.18 (s, 1H), 1.87 – 1.80 (m, 1H), 1.73 (tdd, *J* = 8.3, 6.5, 4.6 Hz, 1H), 1.53 – 1.40 (m, 4H), 1.35 (s, 9H), 0.93 (t, *J*=6.9 Hz, 3H).

^{13}C NMR (126 MHz, $CDCl_3$) ^{13}C NMR (126 MHz, $CDCl_3$) δ 199.9, 156.8, 134.5, 128.2, 125.7, 123.8, 60.12, 35.5, 35.4, 31.3, 30.2, 28.3, 22.4, 14.0.

IR(neat) 3210, 2956, 2923, 2852, 1736, 1606, 1234, 1109, 834, 852, 796 cm^{-1}

LRMS m/z (ESI APCI) calculated for $C_{18}H_{26}O_2$ [M+H] 275.2, found 275.2.

3da *(4-fluorophenyl)(2-(hydroxymethyl)-3-propylcyclopropyl)methanone*

Chemical Formula: $C_{14}H_{17}FO_2$

Y = 69%. Yellow Oil. R_f = 0.16 (4:1 Hexanes:EtOAc).

¹H NMR ¹H NMR (500 MHz, Chloroform-*d*) δ 8.02 (dd, *J* = 8.5, 5.4 Hz, 2H), 7.13 (t, *J* = 8.5 Hz, 2H), 3.94 (dd, *J* = 11.9, 4.7 Hz, 1H), 3.72 (dd, *J* = 12.0, 8.4 Hz, 1H), 2.48 (dd, *J* = 8.4, 5.0 Hz, 1H), 2.14 (s, 1H), 1.82 (p, *J* = 6.2 Hz, 1H), 1.74 (qd, *J* = 8.3, 5.6 Hz, 1H), 1.44 (tt, *J* = 13.9, 7.1 Hz, 4H), 0.92 (t, *J* = 6.8 Hz, 3H).

¹³C NMR (126 MHz, CDCl₃) δ 198.5, 166.8, 164.8, 135.0, 135.0, 130.9, 130.8, 115.9, 115.7, 60.0, 35.6, 35.4, 30.1, 28.4, 22.4, 14.0.

¹⁹F NMR (282 MHz, Chloroform-*d*) δ -104.95 (ddd, *J* = 13.7, 8.5, 5.4 Hz).

IR(neat) 3458, 2958, 2926, 1667, 1599, 1507, 1229, 1155, 1031, 838 cm⁻¹

LRMS m/z (ESI APCI) calculated for $C_{14}H_{17}FO_2$ [M+H] 237.1, found 237.1.

3ea *(2-(hydroxymethyl)-3-propylcyclopropyl)(4-methoxyphenyl)methanone*

MeO

n-Pr

OH

O

Chemical Formula: C$_{15}$H$_{20}$O$_3$

Y = 77%. Pale-Yellow Oil. R_f = 0.06 (4:1 Hexanes:EtOAc).

^{1}H NMR (500 MHz, Chloroform-*d*) δ 7.99 (d, *J* = 8.8 Hz, 2H), 6.95 (d, *J* = 8.8 Hz, 2H), 3.94 (dd, *J* = 12.0, 4.6 Hz, 1H), 3.87 (s, 3H), 3.75 (dd, *J* = 12.1, 8.2 Hz, 1H), 2.49 (dd, *J* = 8.4, 5.0 Hz, 1H), 2.26 (s, 1H), 1.80 (p, *J* = 6.3 Hz, 1H), 1.73 – 1.66 (m, 1H), 1.51 – 1.38 (m, 4H), 0.93 (t, *J* = 6.9 Hz, 3H).

^{13}C NMR (126 MHz, CDCl$_3$) δ 198.7, 131.6, 130.5, 113.9, 60.2, 55.7, 35.5, 35.1, 29.9, 28.0, 22.4, 14.0.

IR(neat) 3436, 2957, 2926, 1655, 1600, 1235, 1170, 1026, 845 cm^{-1}

LRMS m/z (ESI APCI) calculated for C$_{15}$H$_{20}$O$_3$ [M+H] 249.1, found 249.1.

3fa *(2-(hydroxymethyl)-3-propylcyclopropyl)(m-tolyl)methanone*

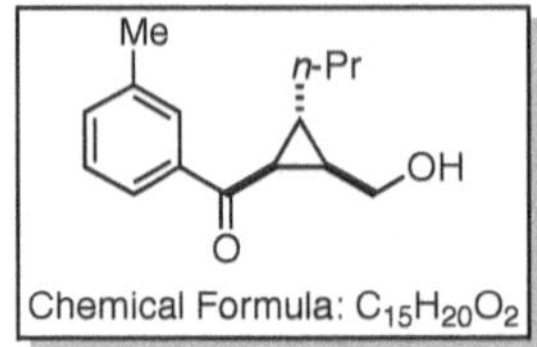

Y = 52%. Pale-yellow oil. R_f = 0.30 (4:1 Hexanes:EtOAc).

¹H NMR (500 MHz, Chloroform-*d*) δ 7.80 (q, *J* = 2.4 Hz, 2H), 7.42 – 7.32 (m, 2H), 3.94 (dd, *J* = 12.0, 4.6 Hz, 1H), 3.76 (dd, *J* = 12.0, 8.3 Hz, 1H), 2.54 (dd, *J* = 8.4, 5.0 Hz, 1H), 2.42 (s, 3H), 2.15 (s, 1H), 1.88 – 1.79 (m, 1H), 1.74 (tdd, *J* = 8.4, 6.5, 4.6 Hz, 1H), 1.51 – 1.38 (m, 4H), 0.97 – 0.89 (m, 3H).

¹³C NMR (126 MHz, CDCl₃) δ 200.3, 138.5, 138.4, 133.6, 128.6, 128.4, 125.3, 59.9, 35.4, 35.3, 30.1, 28.3, 22.2, 21.4, 13.8.

IR(neat) 3445, 2955, 2870, 1664, 1604, 1163, 1054, 1030, 708 cm⁻¹

LRMS m/z (ESI APCI) calculated for $C_{15}H_{20}O_2$ [M+H] 233.2, found 233.2.

3ga *2-(hydroxymethyl)-3-propylcyclopropyl)(3-methoxyphenyl)methanone*

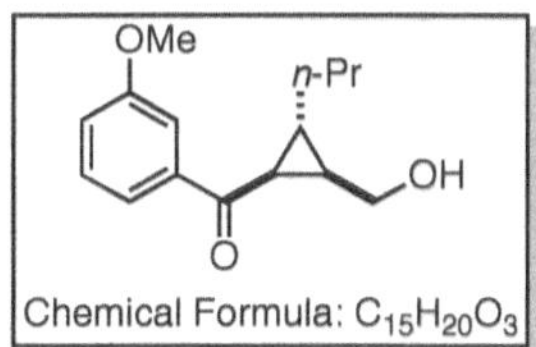

Y = 93%. Yellow oil. R_f = 0.18 (4:1, Hexanes:EtOAc).

¹H NMR (500 MHz, Chloroform-*d*) δ 7.60 (d, *J* = 8.2 Hz, 1H), 7.49 (t, *J* = 2.1 Hz, 1H), 7.38 (t, *J* = 7.9 Hz, 1H), 7.10 (dd, *J* = 8.2, 2.7 Hz, 1H), 3.94 (dd, *J* = 11.9, 4.7 Hz, 1H), 3.85 (s, 3H), 3.75 (dd, *J* = 12.2, 8.1 Hz, 1H), 2.53 (dd, *J* = 8.4, 5.0 Hz, 1H), 2.19 (bs, 1H), 1.86 – 1.79 (m, 1H), 1.78 – 1.70 (m, 1H), 1.53 – 1.37 (m, 4H), 0.96 – 0.90 (m, 3H).

¹³C NMR (126 MHz, CDCl$_3$) δ 200.0, 159.9, 139.9, 129.7, 120.9, 119.4, 112.5, 60.0, 55.6, 35.6, 35.4, 30.3, 28.6, 22.3, 13.9.

IR (neat) 3437, 2957, 2926, 1664, 1586, 1462, 1261, 1034, 778 cm⁻¹

LRMS m/z (ESI APCI) calculated for C$_{15}$H$_{20}$O$_3$ [M+H] 249.1, found 249.1.

3ha *(3-fluorophenyl)-2-(hydroxymethyl)-3-propylcyclopropyl)methanone*

Chemical Formula: $C_{14}H_{17}FO_2$

Y = 54%. Yellow oil. R_f = 0.15 (4:1, Hexanes:EtOAc).

^{1}**H NMR** (500 MHz, Chloroform-*d*) δ 7.79 (dt, *J* = 7.8, 1.2 Hz, 1H), 7.66 (ddd, *J* = 9.5, 2.6, 1.6 Hz, 1H), 7.45 (td, *J* = 8.0, 5.5 Hz, 1H), 7.29 – 7.23 (m, 2H), 3.95 (dd, *J* = 11.9, 4.7 Hz, 1H), 3.73 (dd, *J* = 11.9, 8.4 Hz, 1H), 2.50 (dd, *J* = 8.4, 5.0 Hz, 1H), 1.98 (s, 1H), 1.88 – 1.82 (m, 1H), 1.78 (tdd, *J* = 8.5, 6.6, 4.7 Hz, 1H), 1.52 – 1.38 (m, 4H), 0.98 – 0.89 (m, 3H).

^{13}**C NMR** (126 MHz, CDCl$_3$) δ 198.8 (d, *J* = 2.0 Hz), 164.0, 162.0, 130.4 (d, *J* = 7.7 Hz), 124.0 (d, *J* = 3.1 Hz), 120.0 (d, *J* = 21.6 Hz), 115.0 (d, *J* = 22.3 Hz), 59.9, 35.9, 35.34, 30.4, 28.8, 22.4, 14.0.

^{19}**F NMR** (282 MHz, Chloroform-*d*) δ -111.15 (td, *J* = 9.0, 5.7 Hz).

IR(neat) 3439, 2958, 2925, 1670, 1588, 1443, 1252, 1030, 785 cm^{-1}

LRMS m/z (ESI APCI) calculated for $C_{14}H_{17}FO_2$ [M+H] 237.1, found 237.1.

3ia *(2-fluorophenyl)(2-(hydroxymethyl)-3-propylcyclopropyl)methanone*

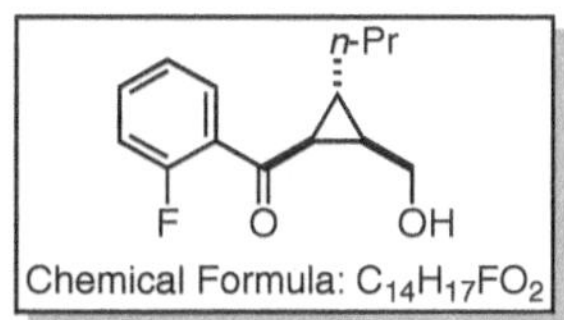

Y = 44%. Pale-yellow Oil. R_f = 0.19 (4:1 Hexanes:EtOAc).

¹H NMR (300 MHz, Chloroform-*d*) δ 7.74 (td, *J* = 7.6, 1.9 Hz, 1H), 7.49 (dddd, *J* = 8.5, 7.1, 5.0, 1.9 Hz, 1H), 7.28 – 7.07 (m, 2H), 3.96 (dd, *J* = 12.0, 4.7 Hz, 1H), 3.78 (t, *J* = 10.1 Hz, 1H), 2.54 (ddd, *J* = 8.1, 5.1, 2.7 Hz, 1H), 1.95 (s, 1H), 1.89 (ddd, *J* = 6.5, 5.1, 1.3 Hz, 1H), 1.77 (tdd, *J* = 8.3, 6.6, 4.8 Hz, 1H), 1.51 – 1.34 (m, 4H), 0.98 – 0.88 (m, 3H).

¹³C NMR (126 MHz, CDCl₃) δ 198.8 (d, *J* = 3.1 Hz), 161.6 (d, *J* = 255.0 Hz), 134.2 (d, *J* = 9.0 Hz), 130.4 (d, *J* = 2.6 Hz), 128.0 (d, *J* = 12.6 Hz), 124.6 (d, *J* = 3.6 Hz), 116.8 (d, *J* = 25.1 Hz), 59.9, 36.3, 35.3, 34.4 (d, *J* = 7.7 Hz), 29.6, 22.3, 14.0.

¹⁹F NMR (282 MHz, CDCl₃) δ -110.33 (dt, *J* = 8.3, 4.0 Hz).

IR(neat) 3213, 2956, 2922, 1653, 1607, 1234, 1109, 1036, 834 cm⁻¹

LRMS m/z (ESI APCI) calculated for C₁₄H₁₆F₂O₂ [M+H] 237.1, found 237.1.

3ja *2-(hydroxymethyl)-3-propylcyclopropyl) (naphthalen-2-yl)methanone*

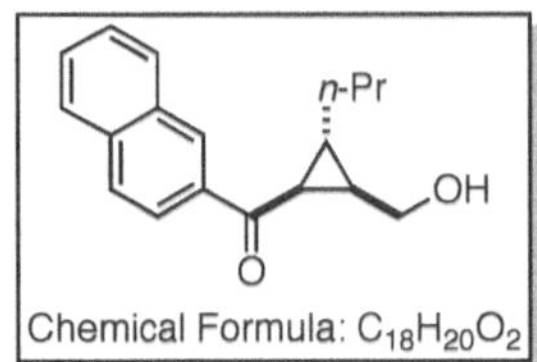

Y = 50%. Pale-yellow solid. R_f = 0.18 (4:1 Hexanes:EtOAc).

^{1}H NMR (500 MHz, Chloroform-*d*) δ 8.55 (d, *J* = 1.7 Hz, 1H), 8.05 (dd, *J* = 8.6, 1.8 Hz, 1H), 7.98 (d, *J* = 8.1 Hz, 1H), 7.89 (dd, *J* = 10.8, 8.3 Hz, 2H), 7.58 (dddd, *J* = 21.9, 8.1, 6.8, 1.3 Hz, 2H), 3.99 (dd, *J* = 12.0, 4.7 Hz, 1H), 3.80 (dd, *J* = 12.0, 8.3 Hz, 1H), 2.71 (dd, *J* = 8.4, 5.0 Hz, 1H), 2.14 (s, 0H), 1.96 – 1.87 (m, 1H), 1.81 (tdd, *J* = 8.3, 6.6, 4.7 Hz, 1H), 1.50 (dtd, *J* = 14.3, 12.8, 11.8, 6.9 Hz, 4H), 0.95 (t, *J* = 7.0 Hz, 3H).

^{13}C NMR (126 MHz, CDCl$_3$) 200.0, 135.9, 135.6, 132.7, 129.8, 129.7, 128.5, 128.5, 127.9, 126.9, 124.1, 60.1, 35.7, 35.5, 30.3, 28.5, 22.4, 14.0.

IR(neat) 3300, 2872, 2857, 1657, 1181, 1123, 1046, 1027, 822, 749 cm^{-1}

LRMS m/z (ESI APCI) calculated for C$_{18}$H$_{20}$O$_2$ [M+H] 269.2, found 269.2.

3ka *1-(2-(hydroxymethyl)-3-propylcyclopropyl)-3-phenylpropan-1-one*

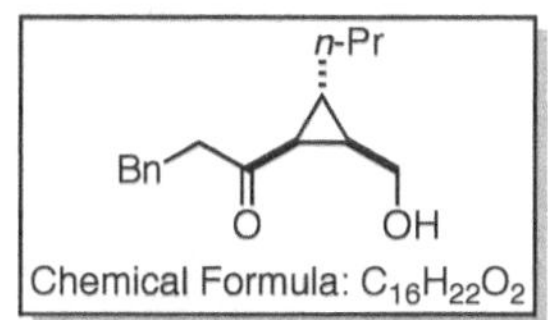

¹H NMR (500 MHz, Chloroform-*d*) δ 7.31 – 7.26 (m, 2H), 7.21 – 7.18 (m, 3H), 3.90 – 3.80 (m, 1H), 3.65 (t, *J* = 10.2 Hz, 1H), 3.02 – 2.88 (m, 4H), 2.01 (s, 1H), 1.83 (dd, *J* = 8.3, 5.0 Hz, 1H), 1.62 (qd, *J* = 6.6, 4.9 Hz, 1H), 1.53 (tdd, *J* = 8.2, 6.6, 4.4 Hz, 1H), 1.42 – 1.28 (m, 3H), 0.89 (t, *J* = 7.2 Hz, 3H).

¹³C NMR (126 MHz, CDCl$_3$) δ 210.2, 141.2, 128.7, 128.5, 126.3, 59.5, 46.3, 35.3, 33.1, 30.2, 28.7, 22.3, 13.9.

IR(neat) 3386, 2958, 2925, 2872, 1689, 1454, 1378, 1034, 733, 700 cm⁻¹

LRMS m/z (ESI APCI) calculated for C$_{16}$H$_{22}$O$_2$ [M+H] 247.3, found 247.3.

3ac *(2-(hydroxymethyl)-3-methylcyclopropyl)(phenyl)methanone*

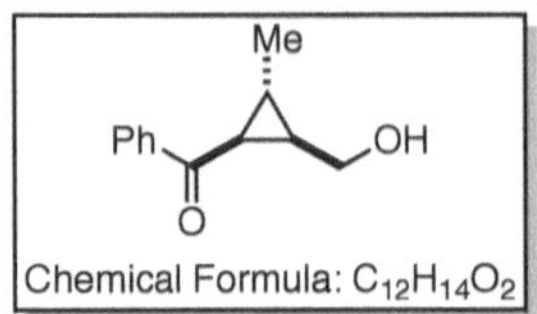

Y = 89%. Light yellow solid. R_f = 0.50 (4:1 Hexanes:EtOAc).

¹H NMR (500 MHz, Chloroform-*d*) δ 8.06 – 7.94 (m, 2H), 7.62 – 7.52 (m, 1H), 7.47 (dd, *J* = 8.3, 7.0 Hz, 2H), 3.96 (dd, *J* = 12.0, 4.7 Hz, 1H), 3.74 (dd, *J* = 12.0, 8.3 Hz, 1H), 2.50 (dd, *J* = 8.4, 5.0 Hz, 1H), 2.17 (s, 1H), 1.84 (h, *J* = 6.0 Hz, 1H), 1.73 (tdd, *J* = 8.3, 6.4, 4.6 Hz, 1H), 1.26 (d, *J* = 6.0 Hz, 4H).

¹³C NMR (126 MHz, CDCl₃) δ 200.3, 138.6, 133.1, 128.7, 128.3, 60.0, 36.5, 31.5, 22.9, 18.3.

IR(neat) 3412, 2955, 2889, 1660, 1598, 1219, 1021, 743, 688 cm⁻¹

LRMS m/z (ESI APCI) calculated for $C_{12}H_{14}O_2$ [M+H] 191.1, found 191.1.

3ad *(2-(hydroxymethyl)-2-methylcyclopropyl)(phenyl)methanone*

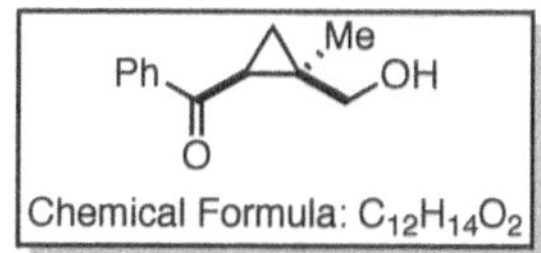

Y = 62% Pale-yellow solid. R_f = 0.19 (4:1 Hexanes:EtOAc)

¹H NMR (500 MHz, Chloroform-*d*) δ 8.02 – 7.97 (m, 2H), 7.59 – 7.54 (m, 1H), 7.47 (dd, *J* = 8.4, 7.0 Hz, 2H), 3.75 (dd, *J* = 11.8, 4.6 Hz, 1H), 3.61 (dd, *J* = 11.8, 5.0 Hz, 1H), 2.54 (dd, *J* = 7.8, 5.7 Hz, 1H), 1.89 (t, *J* = 6.0 Hz, 1H), 1.64 (dd, *J* = 5.7, 4.3 Hz, 1H), 1.42 (s, 3H), 1.10 (dd, *J* = 7.8, 4.3 Hz, 1H).

¹³C NMR (126 MHz, CDCl₃) δ 199.7, 138.6, 133.0, 128.7, 128.3, 64.7, 33.3, 31.5, 22.8, 21.1.

IR(neat) 3439, 2914, 1669, 1269, 1230, 1022, 714

LRMS m/z (ESI APCI) calculated for C₁₂H₁₄O₂ [M+H] 191.1, found 191.1.

3ae *(3-(hydroxymethyl)-2,2-dimethylcyclopropyl)(phenyl)methanone*

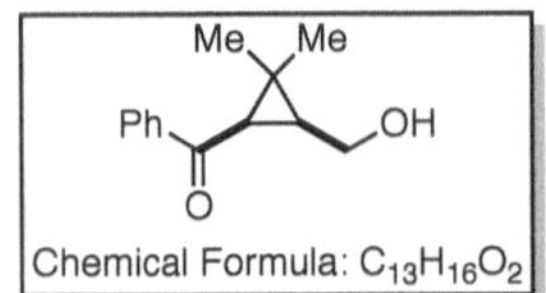

Y = 82% Pale-yellow oil. R_f = 0.18 (4:1 Hexanes:EtOAc)

¹H NMR (500 MHz, Chloroform-*d*) δ 7.98 – 7.89 (m, 2H), 7.61 – 7.54 (m, 1H), 7.48 (dd, *J* = 8.4, 7.0 Hz, 2H), 4.10 – 4.03 (m, 1H), 4.02 – 3.95 (m, 1H), 2.62 (q, *J* = 3.8 Hz, 1H), 2.41 (d, *J* = 7.9 Hz, 1H), 1.66 (ddd, *J* = 9.6, 8.0, 6.9 Hz, 1H), 1.41 (s, 3H), 1.11 (s, 3H).

¹³C NMR (126 MHz, CDCl₃) δ 200.4, 138.8, 133.2, 128.8, 128.3, 59.5, 35.7, 35.6, 28.8, 28.5, 15.4.

IR(neat) 3222, 2911, 1668, 1273, 1220, 692 cm⁻¹

LRMS m/z (ESI APCI) calculated for C₁₃H₁₆O₂ [M+H] 205.1, found 205.1.

3af *(3-(hydroxymethyl)-2-methyl-2-(4-methylpent-3-en-1-yl)cyclopropyl)(phenyl)methanone*

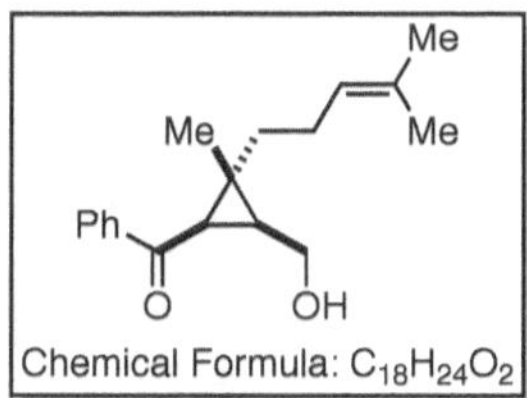

Y = 55%. Colorless oil. R_f = 0.24 (4:1 Hexanes:EtOAc).

¹H NMR (Major) (500 MHz, Chloroform-*d*) δ 7.99 – 7.89 (m, 2H), 7.61 – 7.51 (m, 1H), 7.46 (dd, *J* = 8.5, 7.0 Hz, 2H), 5.13 (tt, *J* = 6.9, 1.5 Hz, 1H), 4.09 – 3.98 (m, 2H), 2.45 (d, *J* = 8.0 Hz, 1H), 2.15 (hept, *J* = 7.5 Hz, 2H), 1.86 (ddd, *J* = 13.4, 9.2, 5.9 Hz, 1H), 1.69 (s, 3H), 1.63 (s, 3H), 1.36 (ddd, *J* = 13.5, 9.7, 6.8 Hz, 1H), 1.12 (s, 3H).

¹³C NMR (Major) (126 MHz, CDCl₃) δ 200.3, 138.9, 134.5, 133.1, 132.4, 128.7, 128.3, 123.8, 123.7, 59.4, 42.9, 35.2, 35.1, 32.5, 25.9, 25.3, 17.9, 12.6.

IR(neat) 3444, 2923, 1725, 1351, 1282, 1261, 1125, 1053, 964, 910, 759, 698 cm⁻¹

LRMS m/z (ESI APCI) calculated for $C_{17}H_{16}O_2$ [M+H] 273.2, found 273.2.

3ag *(2-(1-hydroxyallyl)cyclopropyl)(phenyl)methanone*

Chemical Formula: $C_{13}H_{14}O_2$

Y = 73%. Yellow oil. R_f = 0.22 (4:1 Hexanes:EtOAc).

^{1}H NMR (500 MHz, Chloroform-*d*) δ 8.03 (d, *J* = 7.7 Hz, 2H), 7.56 (t, *J* = 7.4 Hz, 1H), 7.47 (t, *J* = 7.6 Hz, 2H), 6.00 (ddd, *J* = 16.6, 10.4, 5.6 Hz, 1H), 5.30 (d, *J* = 17.2 Hz, 1H), 5.12 (d, *J* = 10.4 Hz, 1H), 4.06 (dd, *J* = 9.0, 5.6 Hz, 1H), 2.81 (td, *J* = 8.1, 5.7 Hz, 1H), 2.16 (s, 1H), 1.75 (p, *J* = 8.4 Hz, 1H), 1.49 (q, *J* = 5.8 Hz, 1H), 1.26 (td, *J* = 7.9, 4.5 Hz, 2H).

^{13}C NMR (126 MHz, CDCl$_3$) δ 199.9, 139.9, 138.6, 133.1, 128.7, 128.4, 114.9, 71.0, 31.1, 23.1, 13.0.

IR(neat) 3407, 2889, 1666, 1391, 1224, 1003, 714, 690 cm^{-1}

LRMS m/z (ESI APCI) calculated for $C_{13}H_{14}O_2$ [M+H] 203.1, found 203.1.

3ah *(2-(1-hydroxyethyl)cyclopropyl)(phenyl)methanone*

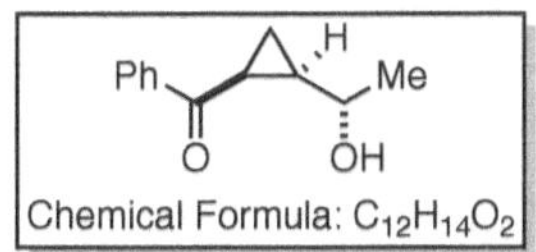

Y = 69%. Off-white solid. R_f = 0.26 (4:1 Hexanes:EtOAc).

¹H NMR (Major) (500 MHz, Chloroform-*d*) δ 8.08 – 8.02 (m, 2H), 7.60 – 7.53 (m, 1H), 7.47 (t, *J* = 7.6 Hz, 2H), 3.74 (dt, *J* = 12.6, 6.4 Hz, 1H), 2.76 (td, *J* = 8.2, 5.7 Hz, 1H), 1.95 (s, 1H), 1.69 (qd, *J* = 8.7, 6.9 Hz, 1H), 1.43 – 1.35 (m, 1H), 1.33 (d, *J* = 6.3 Hz, 3H), 1.25 (td, *J* = 8.2, 4.4 Hz, 1H).

¹³C NMR (Major) (126 MHz, CDCl₃) δ 199.9, 138.5, 133.1, 128.7, 128.4, 66.6, 33.0, 23.4, 23.2, 13.2.

IR(neat) 3497, 2965, 1666, 1599, 1390, 1210, 999, 699, 690 cm⁻¹

LRMS m/z (ESI APCI) calculated for $C_{12}H_{14}O_2$ [M+H] 191.1, found 191.1.

3ai *(2-(hydroxy(phenyl)methyl)cyclopropyl)(phenyl)methanone*

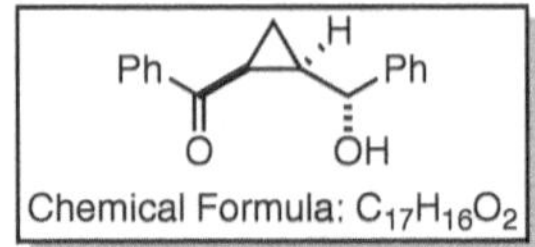

Y = 62%. Light yellow solid. R_f = 0.54 (4:1 Hexanes:EtOAc).

¹H NMR (500 MHz, Chloroform-*d*) δ 8.17 – 8.06 (m, 2H), 7.61 (t, *J* = 7.4 Hz, 1H), 7.56 – 7.47 (m, 4H), 7.39 (t, *J* = 7.5 Hz, 2H), 7.32 (t, *J* = 7.3 Hz, 1H), 4.64 (d, *J* = 9.4 Hz, 1H), 2.93 (td, *J* = 8.1, 5.7 Hz, 1H), 2.26 (s, 1H), 2.01 (qd, *J* = 8.7, 6.9 Hz, 1H), 1.64 (dt, *J* = 6.8, 5.0 Hz, 1H), 1.26 (dt, *J* = 8.2, 4.1 Hz, 1H).

¹³C NMR (126 MHz, CDCl₃) δ 199.6, 144.0, 138.6, 133.0, 128.7, 128.6, 128.4, 127.7, 126.0, 72.1, 33.0, 23.8, 13.4.

IR(neat) 3495, 2944, 1669, 1591, 1390, 1223, 1210, 1010, 699 cm⁻¹

LRMS m/z (ESI APCI) calculated for $C_{17}H_{16}O_2$ [M+H] 253.1, found 253.1.

3aj *(2-(1-hydroxyethyl)-3-methylcyclopropyl)(phenyl)methanone*

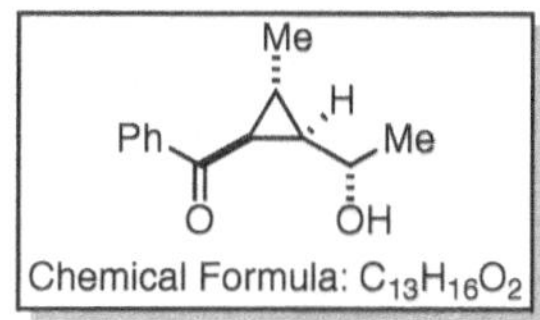

Y = 88% Pale-yellow oil. R_f = 0.19 (4:1 Hexanes:EtOAc)

^{1}H NMR (500 MHz, Chloroform-*d*) δ 8.05 – 7.98 (m, 2H), 7.60 – 7.53 (m, 1H), 7.51 – 7.43 (m, 2H), 3.87 (dq, *J* = 8.8, 6.3 Hz, 1H), 2.47 (dd, *J* = 8.4, 5.1 Hz, 1H), 2.23 (s, 1H), 1.74 (td, *J* = 6.3, 5.2 Hz, 1H), 1.50 (td, *J* = 8.6, 6.4 Hz, 1H), 1.32 (d, *J* = 6.3 Hz, 3H), 1.24 (d, *J* = 6.1 Hz, 3H).

^{13}C NMR (126 MHz, CDCl$_3$) δ 200.2, 138.6, 133.0, 128.7, 128.7, 128.7, 128.2, 66.0, 42.2, 32.0, 23.2, 23.1, 18.2.

IR(neat) 3214, 2975, 1735, 1602, 1224, 1115, 1062, 905, 732, 715 cm^{-1}

LRMS m/z (ESI APCI) calculated $C_{13}H_{16}O_2$ [M+H] 205.1, found 205.1.

3ak *(2-(hydroxy(phenyl)methyl)-3-propylcyclopropyl)(phenyl)methanone*

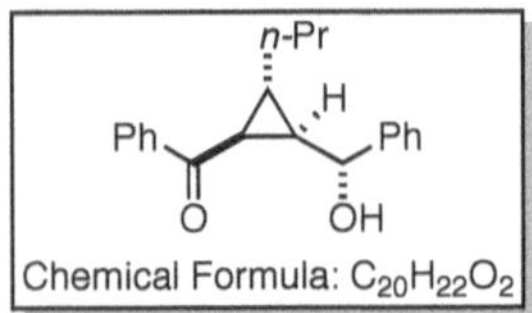

Y = 75%. Off-white solid. R_f = 0.34 (4:1 Hexanes:EtOAc).

¹H NMR (500 MHz, Chloroform-*d*) δ 8.09 – 8.02 (m, 2H), 7.61 – 7.54 (m, 1H), 7.52 – 7.42 (m, 4H), 7.36 (t, *J* = 7.6 Hz, 2H), 7.31 – 7.24 (m, 1H), 4.77 (dd, *J* = 9.2, 3.7 Hz, 1H), 2.66 (dd, *J* = 8.4, 5.1 Hz, 1H), 2.28 (d, *J* = 3.8 Hz, 1H), 1.99 (qd, *J* = 6.6, 5.0 Hz, 1H), 1.81 (td, *J* = 8.9, 6.5 Hz, 1H), 1.48 – 1.37 (m, 1H), 1.31 – 1.26 (m, 1H), 1.22 (ddtd, *J* = 12.6, 8.1, 6.7, 6.3, 4.4 Hz, 2H), 0.77 (t, *J* = 7.2 Hz, 3H).

¹³C NMR (126 MHz, CDCl$_3$) δ 199.8, 144.1, 138.7, 132.9, 128.7, 128.6, 127.6, 126.0, 71.6, 41.3, 35.2, 31.2, 28.6, 22.1, 13.8.

IR(neat) 3437, 2958, 2923, 1666, 1587, 1442, 1251, 1032, 908, 730 cm⁻¹

LRMS m/z (ESI APCI) calculated for C$_{20}$H$_{22}$O$_2$ [M+H] 295.2, found 295.2.

3al *(2-(cyclohexyl(hydroxy)methyl)-3-propylcyclopropyl)(phenyl)methanone*

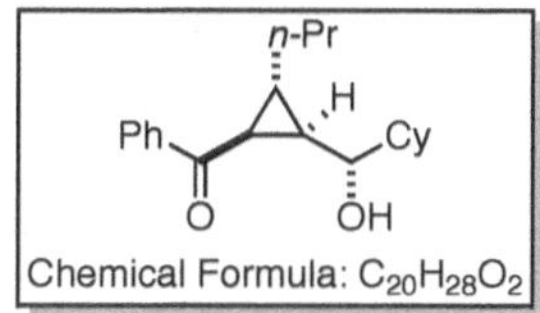

Y = 95%. Off-white solid. R_f = 0.54 (4:1 Hexanes:EtOAc).

^{1}H NMR (500 MHz, Chloroform-*d*) δ 8.06 – 7.97 (m, 2H), 7.54 (t, *J* = 7.5 Hz, 1H), 7.45 (t, *J* = 7.6 Hz, 2H), 3.41 (dd, *J* = 9.2, 6.6 Hz, 1H), 2.49 (dd, *J* = 8.7, 5.0 Hz, 1H), 1.99 – 1.86 (m, 2H), 1.86 – 1.80 (m, 1H), 1.80 – 1.71 (m, 3H), 1.62 (dtq, *J* = 12.8, 10.2, 3.8, 2.9 Hz, 3H), 1.43 (dt, *J* = 14.8, 7.4 Hz, 3H), 1.32 – 1.15 (m, 4H), 1.15 – 1.00 (m, 2H), 0.92 (t, *J* = 7.3 Hz, 3H).

^{13}C NMR (126 MHz, CDCl$_3$) δ 200.3, 138.8, 132.8, 128.6, 128.3, 73.7, 44.4, 38.8, 35.4, 29.3, 29.0, 26.7, 26.4, 26.3, 22.2, 14.1.

IR(neat) 3438, 2922, 2851, 1656, 1451, 1232, 1020, 711, 685, 660 cm^{-1}

LRMS m/z (ESI APCI) calculated for C$_{20}$H$_{28}$O$_2$ [M+H] 301.2, found 301.2.

3an *(2-hydroxybicyclo[6.1.0]nonan-9-yl)(phenyl)methanone*

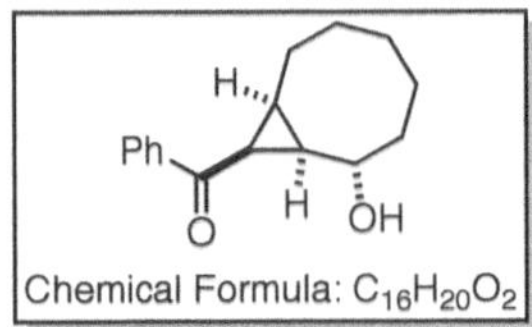

Y = 85%. Colorless oil. R_f = 0.05 (4:1 Hexanes:EtOAc)

¹H NMR (500 MHz, Chloroform-*d*) δ 8.10 – 7.98 (m, 2H), 7.57 (t, *J* = 7.4 Hz, 1H), 7.47 (t, *J* = 7.6 Hz, 2H), 4.13 (td, *J* = 10.9, 4.6 Hz, 1H), 3.14 (s, 1H), 2.58 (dd, *J* = 9.6, 7.7 Hz, 1H), 1.91 (tt, *J* = 12.0, 4.3 Hz, 1H), 1.77 (tdd, *J* = 17.0, 8.4, 4.2 Hz, 3H), 1.73 – 1.65 (m, 2H), 1.59 (ddt, *J* = 15.3, 12.2, 4.3 Hz, 1H), 1.49 (tdd, *J* = 13.6, 10.4, 4.4 Hz, 1H), 1.45 – 1.30 (m, 2H), 1.21 (tdd, *J* = 14.3, 8.0, 3.6 Hz, 2H).

¹³C NMR (126 MHz, CDCl₃) δ 200.3, 138.4, 133.4, 128.7, 128.6, 67.7, 36.7, 30.4, 29.8, 27.0, 26.6, 26.2, 25.1, 24.5.

IR(neat) 3446, 2924, 2854, 1660, 1448, 1394, 1212, 1050, 1011, 956, 718 cm⁻¹

LRMS m/z (ESI APCI) calculated for C₁₆H₂₀O₂ [M+H] 245.3, found 245.3.

A2.4 Mechanistic Experiments

Deuterated allylic alcohol–Retention of stereochemistry at the alkene

N-enoxyphthalimide **1a** (0.12 mmol), catalyst [Cp*^{CF3}RhCl$_2$]$_2$ (5 mol%, 0.006 mmol, 4.4 mg), and KOPiv (2 equiv., 0.24 mmol, 33.6 mg) were weighed in a 1-dram vial with a magnetic stirbar. Cooled TFE (0.2 M, 600 μL) was added followed by allylic alcohol **2a-d$_1$** (1.2 equiv., 0.144 mmol). The vial was sealed with a screw-cap and placed in an aluminium block cooled to 0 °C surrounded by ice in an insulated box and stirred for 16 hours. Upon completion judged by TLC, TFE was removed by rotary evaporation and the residue was taken up in EtOAc and filtered through a silica plug flushing with EtOAc. The filtrate was concentrated to ~1mL and transferred to a 1.5-dram vial where the solution was partitioned with the addition of 10% NaOH solution. The aqueous layer was extracted three times with EtOAc and the combined organic extracts were filtered through a pad of celite® and Na$_2$SO$_4$ then concentrated. The residue was purified by flash chromatography (Hexane:EtOAc, 19:1→9:1→4:1) to afford the cyclopropane product.

3aa' *2-(hydroxymethyl)-3-propylcyclopropyl-2-d)(phenyl)methanone*

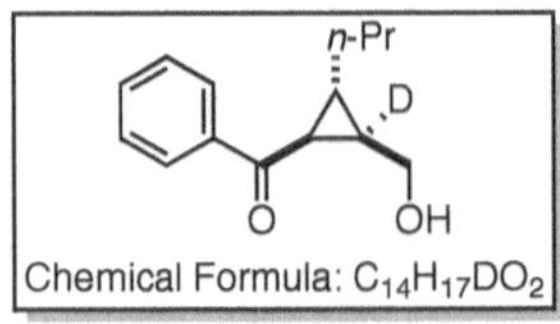

Y = 82%. Colorless Oil. R_f = 0.22 (4:1 Hexanes:EtOAc).

^{1}**H NMR** (500 MHz, Chloroform-*d*) δ 8.05 – 7.97 (m, 2H), 7.62 – 7.55 (m, 1H), 7.49 (dd, *J* = 8.4, 7.0 Hz, 2H), 3.96 (d, *J* = 12.0 Hz, 1H), 3.77 (d, *J* = 12.0 Hz, 1H), 2.56 (d, *J* = 5.0 Hz, 1H), 2.21 (s, 1H), 1.86 (q, *J* = 6.2 Hz, 1H), 1.48 (ttd, *J* = 11.8, 5.8, 5.1, 2.5 Hz, 4H), 0.99 – 0.92 (m, 3H).

^{13}**C NMR** (126 MHz, CDCl$_3$) δ 199.6, 144.0, 138.6, 133.0, 128.7, 128.6, 128.4, 127.7, 126.0, 72.1, 33.0, 23.8, 13.4.

IR(neat) 3458, 2921, 1657, 1450, 1228, 1020, 699 cm^{-1}

LRMS m/z (ESI APCI) calculated for C$_{17}$H$_{16}$O$_2$ [M+H] 220.1, found 220.1.

The assignment of the major diastereomer for the substrates presented in this work was determined by analogy of this result. When **2a-d₁** is subjected to the reaction conditions, the resulting cyclopropane **3aa'** is characterized by coupling constants of the α-hydrogen to the phenyl ketone (highlighted in red) that gives a doublet in the ¹H-NMR spectrum. Assuming retention of stereochemistry at the alkene, if the hydroxymethyl substituent is *trans* to the ketone, the blue proton from the alkene should be *cis* to the ketone and give a large J value. Alternatively, if the hydroxymethyl substituent is *cis* to the ketone, the blue proton from the alkene should be *trans* to the ketone and give a small J value. We observe the doublet of the α-hydrogen to have a J value of 5.0 Hz indicating formation of the diastereomer in the highlighted box.

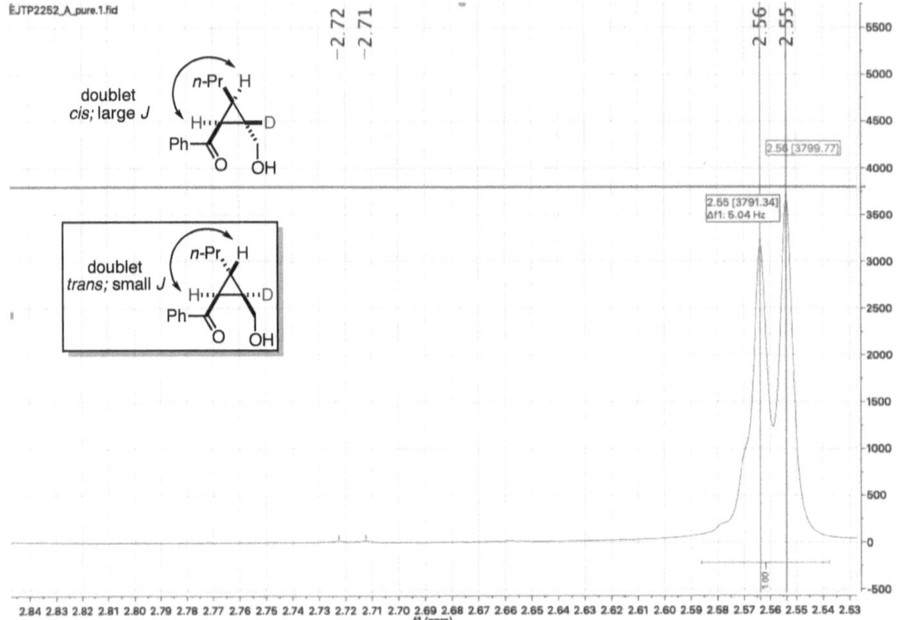

EJTP2252_A_pure.1.fid
doublet
cis; large J
n-Pr
H
H
D
Ph
O
OH
doublet
trans; small J
n-Pr
H
H
D
Ph
O
OH
2.72
2.71
2.56
2.55
2.56 [3799.77]
2.55 [3791.34]
Δf1: 5.04 Hz

N-enoxyphthalimide **1a** (0.12 mmol), catalyst [Cp*CF3RhCl₂]₂ (5 mol%, 0.006 mmol, 4.4 mg), and KOPiv (2 equiv., 0.24 mmol, 33.6 mg) were weighed in a 1-dram vial with a magnetic stirbar. Cooled TFE (0.2 M, 600 µL) was added. The vial was sealed with a screw-cap and placed in an aluminium block cooled to 0 °C surrounded by ice in an insulated box and stirred for 3 hours. TFE-d_1 was removed by rotary evaporation and the residue was taken up in EtOAc and filtered through a silica plug flushing with EtOAc. The filtrate was concentrated to ~1mL and transferred to a 1.5-dram vial where the solution was partitioned with the addition of 10% NaOH solution. The aqueous layer was extracted three times with EtOAc and the combined organic extracts were filtered through a pad of celite® and Na₂SO₄ then concentrated. The residue was purified by flash chromatography (Hexane:EtOAc, 19:1→9:1→4:1) to re-isolate the starting material.

*With homoallylic alcohol **4a***

N-enoxyphthalimide **1a** (0.12 mmol), catalyst $[Cp^{*CF3}RhCl_2]_2$ (5 mol%, 0.006 mmol, 4.4 mg), and KOPiv (2 equiv., 0.24 mmol, 33.6 mg) were weighed in a 1-dram vial with a magnetic stirbar. Cooled TFE (0.2 M, 600 μL) was added followed by homoallylic alcohol **4a** (1.2 equiv., 0.144 mmol). The vial was sealed with a screw-cap and placed in an aluminium block cooled to 0 °C surrounded by ice in an insulated box and stirred for 16 hours. Upon completion judged by TLC, TFE was removed by rotary evaporation and the residue was taken up in EtOAc and filtered through a silica plug flushing with EtOAc. The filtrate was concentrated to ~1mL and transferred to a 1.5-dram vial where the solution was partitioned with the addition of 10% NaOH solution. The aqueous layer was extracted three times with EtOAc and the combined organic extracts were filtered through a pad of celite® and Na_2SO_4 then concentrated. Yield and diastereoselectivity were determined by crude ^{1}H NMR.

With bis-homoallylic alcohol

N-enoxyphthalimide **1a** (0.12 mmol), catalyst $[Cp*^{CF3}RhCl_2]_2$ (5 mol%, 0.006 mmol, 4.4 mg), and KOPiv (2 equiv., 0.24 mmol, 33.6 mg) were weighed in a 1-dram vial with a magnetic stirbar. Cooled TFE (0.2 M, 600 μL) was added followed by *bis*-homoallylic alcohol **6a** (1.2 equiv., 0.144 mmol). The vial was sealed with a screw-cap and placed in an aluminium block cooled to 0 °C surrounded by ice in an insulated box and stirred for 16 hours. Upon completion judged by TLC, TFE was removed by rotary evaporation and the residue was taken up in EtOAc and filtered through a silica plug flushing with EtOAc. The filtrate was concentrated to ~1mL and transferred to a 1.5-dram vial where the solution was partitioned with the addition of 10% NaOH solution. The aqueous layer was extracted three times with EtOAc and the combined organic extracts were filtered through a pad of celite® and Na_2SO_4 then concentrated. Yield and diastereoselectivity were determined by crude ^{1}H NMR.

*With allylic ether **6a***

N-enoxyphthalimide **1a** (0.12 mmol), catalyst [Cp*CF3RhCl2]2 (5 mol%, 0.006 mmol, 4.4 mg), and KOPiv (2 equiv., 0.24 mmol, 33.6 mg) were weighed in a 1-dram vial with a magnetic stirbar. Cooled TFE (0.2 M, 600 μL) was added followed by allylic ether **8a** (1.2 equiv., 0.144 mmol). The vial was sealed with a screw-cap and placed in an aluminium block cooled to 0 °C surrounded by ice in an insulated box and stirred for 16 hours. Upon completion judged by TLC, TFE was removed by rotary evaporation and the residue was taken up in EtOAc and filtered through a silica plug flushing with EtOAc. The filtrate was concentrated to ~1mL and transferred to a 1.5-dram vial where the solution was partitioned with the addition of 10% NaOH solution. The aqueous layer was extracted three times with EtOAc and the combined organic extracts were filtered through a pad of celite® and Na_2SO_4 then concentrated. Yield and diastereoselectivity were determined by crude ^{1}H NMR.

With allylic carboxylic acid

N-enoxyphthalimide (0.12 mmol), catalyst [Cp*^{CF3}RhCl$_2$]$_2$ (5 mol%, 0.006 mmol, 4.4 mg), and KOPiv (2 equiv., 0.24 mmol, 33.6 mg) were weighed in a 1-dram vial with a magnetic stirbar. Cooled TFE (0.2 M, 600 μL) was added followed by allylic carboxylic acid (1.2 equiv., 0.144 mmol). The vial was sealed with a screw-cap and placed in an aluminium block cooled to 0 °C surrounded by ice in an insulated box and stirred for 16 hours. Upon completion judged by TLC, TFE was removed by rotary evaporation and the residue was taken up in EtOAc and filtered through a silica plug flushing with EtOAc. The filtrate was concentrated to ~1mL and transferred to a 1.5-dram vial where the solution was partitioned with the addition of 10% NaOH solution. The aqueous layer was extracted three times with EtOAc and the combined organic extracts were filtered through a pad of celite® and Na$_2$SO$_4$ then concentrated. Yield and diastereoselectivity were determined by crude ^{1}H NMR.

With allylic amine

N-enoxyphthalimide (0.12 mmol), catalyst [Cp*CF3RhCl$_2$]$_2$ (5 mol%, 0.006 mmol, 4.4 mg), and KOPiv (2 equiv., 0.24 mmol, 33.6 mg) were weighed in a 1-dram vial with a magnetic stirbar. Cooled TFE (0.2 M, 600 μL) was added followed by allylic amine (1.2 equiv., 0.144 mmol). The vial was sealed with a screw-cap and placed in an aluminium block cooled to 0 °C surrounded by ice in an insulated box and stirred for 16 hours. Upon completion judged by TLC, TFE was removed by rotary evaporation and the residue was taken up in EtOAc and filtered through a silica plug flushing with EtOAc. The filtrate was concentrated to ~1mL and transferred to a 1.5-dram vial where the solution was partitioned with the addition of 10% NaOH solution. The aqueous layer was extracted three times with EtOAc and the combined organic extracts were filtered through a pad of celite® and Na$_2$SO$_4$ then concentrated. The crude residue was purified by flash chromatography (Hexane:EtOAc, 19:1→9:1→4:1) to afford the cyclopropane product.

9ac *N-((2-benzoyl-3-propylcyclopropyl)methyl)-4-methylbenzenesulfonamide*

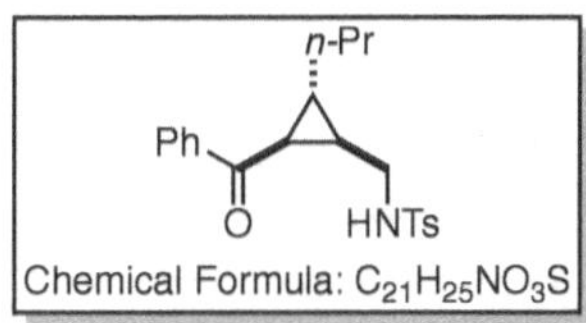

Y = 77%. Off-White Solid. R_f = 0.21 (4:1 Hexanes:EtOAc).

¹H NMR (500 MHz, Chloroform-*d*) δ 7.91 (dd, *J* = 8.3, 1.3 Hz, 2H), 7.66 (d, *J* = 8.2 Hz, 2H), 7.60 – 7.53 (m, 1H), 7.46 (t, *J* = 7.7 Hz, 2H), 7.25 (d, *J* = 8.0 Hz, 2H), 4.65 (t, *J* = 6.4 Hz, 1H), 3.37 (ddd, *J* = 14.0, 6.9, 5.6 Hz, 1H), 3.11 (ddd, *J* = 14.3, 8.8, 5.9 Hz, 1H), 2.49 (dd, *J* = 8.4, 5.0 Hz, 1H), 2.40 (s, 3H), 1.71 (tt, *J* = 8.6, 6.0 Hz, 1H), 1.67 – 1.60 (m, 1H), 1.42 – 1.30 (m, 4H), 0.94 – 0.85 (m, 3H).

¹³C NMR (126 MHz, CDCl₃) δ 199.3, 143.4, 134.5, 133.2, 129.8, 128.8, 128.2, 127.2, 123.8, 41.3, 35.1, 33.0, 29.8, 29.7, 22.3, 21.7, 13.9.

IR(neat) 3520, 3189, 3061, 1666, 1602, 1373, 1305, 1239, 1159, 1049, 711, 647.

LRMS m/z (ESI APCI) calculated for $C_{21}H_{25}O_3S$ [M+H] 372.1, found 372.1.

With NaH

N-enoxyphthalimide **1a** (0.12 mmol) and catalyst [Cp*CF3RhCl$_2$]$_2$ (5 mol%, 0.006

mmol, 4.4 mg), were weighed in a 1-dram vial with a magnetic stirbar. Cooled TFE

(0.2 M, 600 μL) was added followed by allylic alcohol **6a** (1.2 equiv., 0.144 mmol). The

vial was sealed with a screw-cap and placed in an aluminium block cooled to 0 °C

surrounded by ice in an insulated box and stirred for 2 mins. The vial was removed and

NaH (2 equiv., 0.24 mmol) was added and vigorous bubbling occurred. The vial was

placed back in the cooling block at 0 °C and stirred for 16 hours. Upon completion

judged by TLC, TFE was removed by rotary evaporation and the residue was taken up

in EtOAc and filtered through a silica plug flushing with EtOAc. The filtrate was

concentrated to ~1mL and transferred to a 1.5-dram vial where the solution was

partitioned with the addition of 10% NaOH solution. The aqueous layer was extracted

three times with EtOAc and the combined organic extracts were filtered through a pad

of celite® and Na$_2$SO$_4$ then concentrated. Yield and diastereoselectivity were

determined by crude ^{1}H NMR.

Isolation of off-cycle intermediate

N-enoxyphthalimide **1a** (0.75 mmol) and KOPiv (1 equiv., 0.75 mmol) were weighed in a 1-dram vial with a magnetic stirbar. THF (0.2 M, 3.770 mL) was added followed by allylic alcohol **2b** (1.2 equiv., 0.91 mmol). The vial was sealed with a screw-cap and stirred for 16 hours at room temperature. THF was removed by rotary evaporation and the crude residue was purified by flash chromatography (Hexane:EtOAc, 19:1) to afford the dioxazoline product.

10ac *(E)-but-2-en-1-yl 2-(5-methyl-5-phenyl-1,4,2-dioxazol-3-yl)benzoate*

Chemical Formula: $C_{20}H_{19}NO_4$

Y=38% Colorless Oil. R_f = 0.64 (4:1 Hexanes:EtOAc).

^{1}H NMR (500 MHz, Chloroform-*d*) δ 7.80 – 7.74 (m, 1H), 7.71 – 7.65 (m, 1H), 7.62 – 7.58 (m, 2H), 7.56 – 7.50 (m, 2H), 7.44 – 7.37 (m, 3H), 5.81 (dqt, *J* = 15.3, 6.4, 1.2 Hz, 1H), 5.59 (dtq, *J* = 14.8, 6.5, 1.6 Hz, 1H), 4.68 (ddt, *J* = 12.3, 6.5, 1.2 Hz, 1H), 4.61 (ddt, *J* = 12.2, 6.5, 1.1 Hz, 1H), 2.01 (s, 3H), 1.71 (dq, *J* = 6.5, 1.2 Hz, 3H).

^{13}C NMR (126 MHz, CDCl$_3$) δ 166.96, 158.13, 140.15, 132.41, 132.01, 131.28, 131.24, 129.99, 129.63, 129.30, 128.57, 125.22, 124.85, 122.70, 116.05, 66.62, 25.70, 17.98.

IR(neat) 2973, 1726, 1282, 1261, 1121, 910, 759, 697 cm^{-1}

LRMS m/z (ESI APCI) calculated for $C_{20}H_{19}NO_4$ [M+H] 338.1, found 338.1

*Compatibility of **10ac** with the cyclopropanation reaction conditions*

Dioxazoline **10ac** (0.12 mmol), catalyst [Cp*^{CF3}RhCl$_2$]$_2$ (5 mol%, 0.006 mmol, 4.4 mg), and KOPiv (2 equiv., 0.24 mmol, 33.6 mg) were weighed in a 1-dram vial with a magnetic stirbar. Cooled (or room temperature) TFE (0.2 M, 600 μL) was added. The vial was sealed with a screw-cap and placed in an aluminium block cooled to 0 °C surrounded by ice in an insulated box (or placed on a stir plate at room temperature) and stirred for 16 hours. TFE was removed by rotary evaporation and the residue was taken up in EtOAc and filtered through a silica plug flushing with EtOAc. The filtrate was concentrated to ~1mL and transferred to a 1.5-dram vial where the solution was partitioned with the addition of H$_2$O. The aqueous layer was extracted three times with EtOAc and the combined organic extracts were filtered through a pad of celite® and Na$_2$SO$_4$ then concentrated. Yield and diastereoselectivity were determined by crude ^{1}H-NMR.

N-enoxyphthalimide **1aa'** (0.06 mmol) and KOPiv (2 equiv., 0.12 mmol) were weighed

in a 1-dram vial with a magnetic stirbar. TFE-d_3 (0.1 M, 600 mL) was added followed

by allylic alcohol **2c** (1.7 equiv., 0.10 mmol). The vial was sealed with a screw-cap and

stirred briefly and the solution transferred to an NMR tube and injected in the

spectrometer set to 273 K.

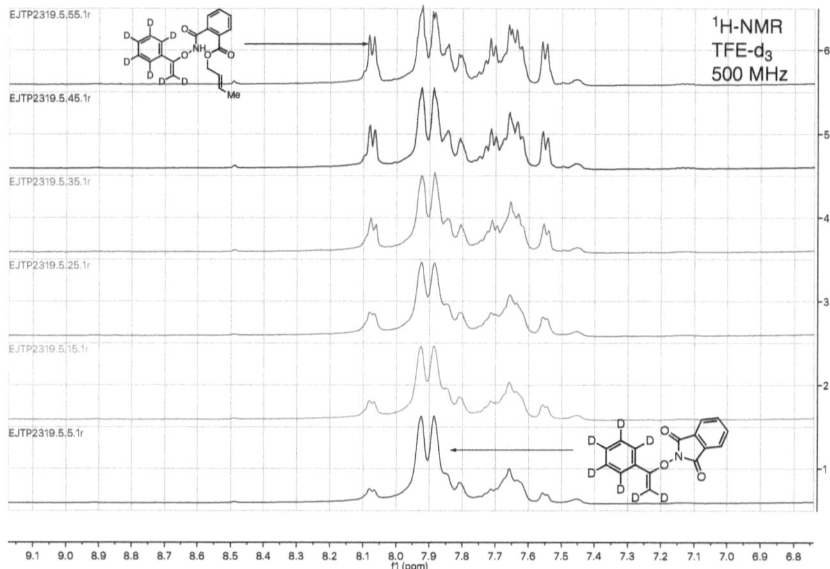

EJTP2319.5.55.1r
EJTP2319.5.45.1r
EJTP2319.5.35.1r
EJTP2319.5.25.1r
EJTP2319.5.15.1r
EJTP2319.5.5.1r
1H-NMR
TFE-d3
500 MHz

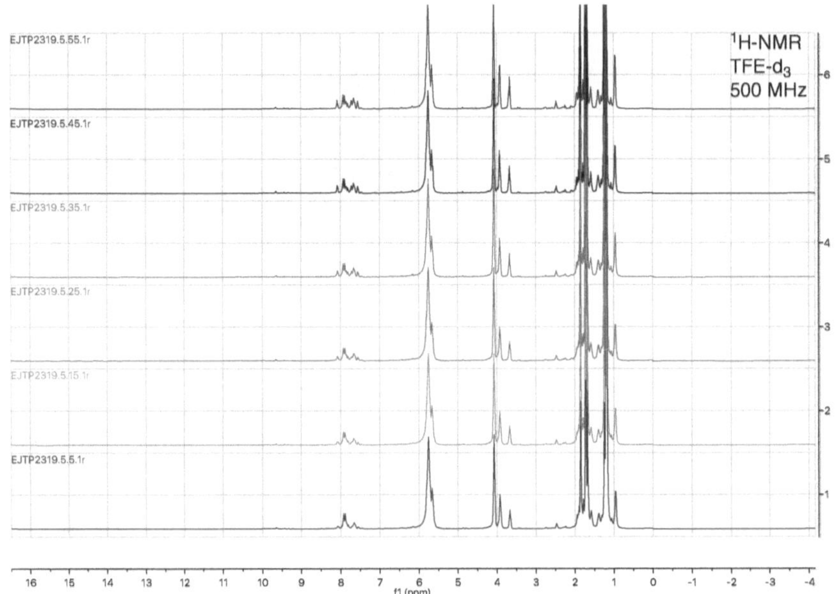

EJTP2319.5.55.1r
EJTP2319.5.45.1r
EJTP2319.5.35.1r
EJTP2319.5.25.1r
EJTP2319.5.15.1r
EJTP2319.5.5.1r
1H-NMR
TFE-d3
500 MHz
f1 (ppm)

A2.5 Model for Diastereoselectivity

A2.6 X-Ray Data

Single crystal X-ray diffraction. Data for all compounds was collected on an Agilent
SuperNova diffractometer using mirror-monochromated Cu Kα radiation. Data
collection, integration, scaling (ABSPACK) and absorption correction (face-indexed
Gaussian integration[12] or numeric analytical methods[13]) were performed in
CrysAlisPro.[14] Structure solution was performed using ShelXT[15]. Subsequent
refinement was performed by full-matrix least-squares on F^2 in ShelXL.[16] Olex2[17] was
used for viewing and to prepare CIF files. PLATON[18] was used extensively for
CheckCIF. ORTEP graphics were prepared in CrystalMaker.[19] Thermal ellipsoids are
rendered at the 50% probability level.

A solution of EJTP2213_B_pure in $CHCl_3$/hexanes was slowly evaporated to afford
long, colorless needles. Part of a crystal (.46 x .06 x .04 mm) was separated carefully,
mounted on a glass fiber with Paratone oil, and cooled to 100 K on the diffractometer.
Complete data were collected to 0.8 Å. 12900 reflections were collected (2666 unique,
2373 observed) with R(int) 5.9% and R(sigma) 4.2% after Gaussian absorption and
beam profile correction (maximum correction factor 1.46).

The space group was assigned tentatively as I2/a based on the systematic absences. Using ShelXT, the structure solved readily in I2/a with 1 molecule in the asymmetric unit. All non-H atoms were located in the initial solution and refined anisotropically with no restraints. The O-H hydrogen was located in a difference map and refined with unrestrained coordinates and isotropic ADP. C-H hydrogens were placed in calculated positions and refined with riding coordinates and ADPs.

The final refinement (2666 data, 0 restraints, 176 parameters) converged with R_1 ($F_o >$ $4\sigma(F_o)$) = 4.6%, wR_2 = 12.1%, S = 1.04. The largest Fourier features were 0.25 and - 0.20 e$^-$ A^{-3}.

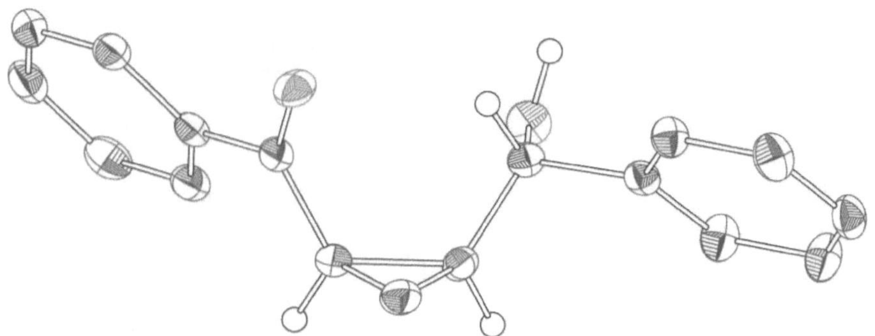

Molecular structure of EJTP2213_B_pure. The crystal is centrosymmetric and thus contains both enantiomers.

Compound	EJTP2213_B_pure

Formula	$C_{17}H_{16}O_2$
MW	252.30
Space group	I2/a
a (Å)	20.0490(6)
b (Å)	5.46156(14)
c (Å)	25.6163(8)
a (°)	90
β (°)	107.386(3)
γ (°)	90
V (Å^3)	2676.80(14)
Z	8
ρ_{calc} (g cm^{-3})	1.252
T (K)	100
λ (Å)	1.54184
$2\theta_{min}$, $2\theta_{max}$	7, 146
Nref	12900
R(int), R(σ)	.0591, .0416
μ(mm^{-1})	0.642
Size (mm)	.46 x .06 x .04
T_{max} / T_{min}	1.46
Data	2666
Restraints	0
Parameters	176
R_1(obs)	0.0458
wR_2(all)	0.1212
S	1.036
Peak, hole (e$^-$ Å^{-3})	0.25, -0.20

A2.7 NMR Spectra

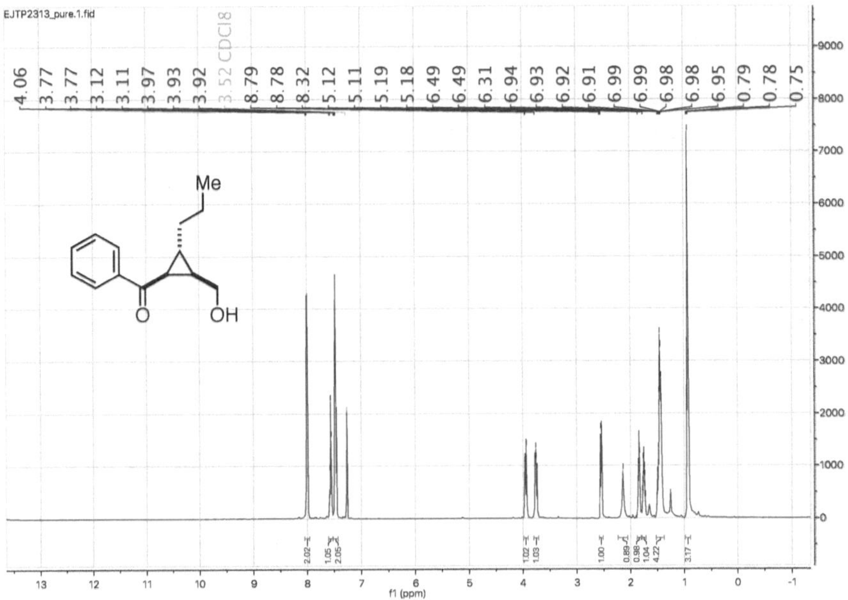

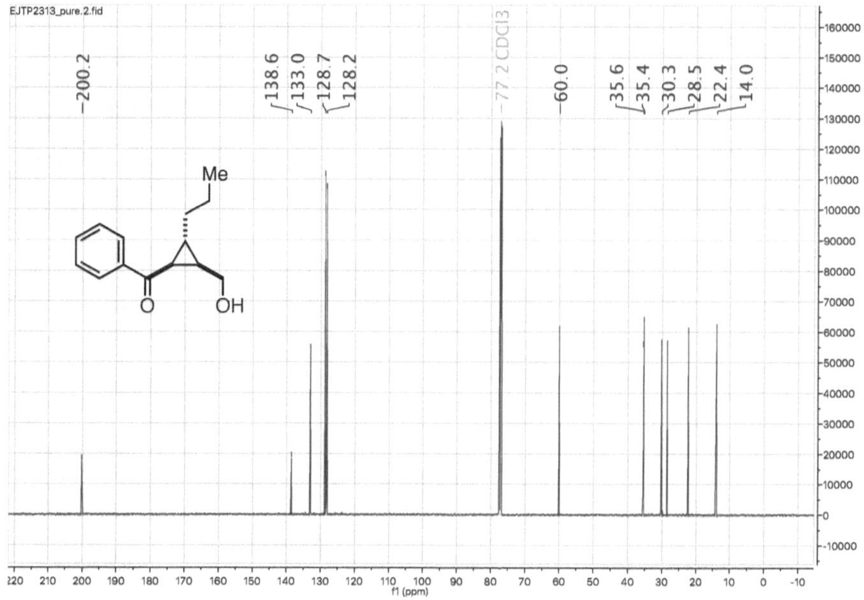

EJTP2313_pure.2.fid
200.2
138.6
133.0
128.7
128.2
77.2 CDCl3
60.0
35.6
35.4
30.3
28.5
22.4
14.0
Me
OH
O

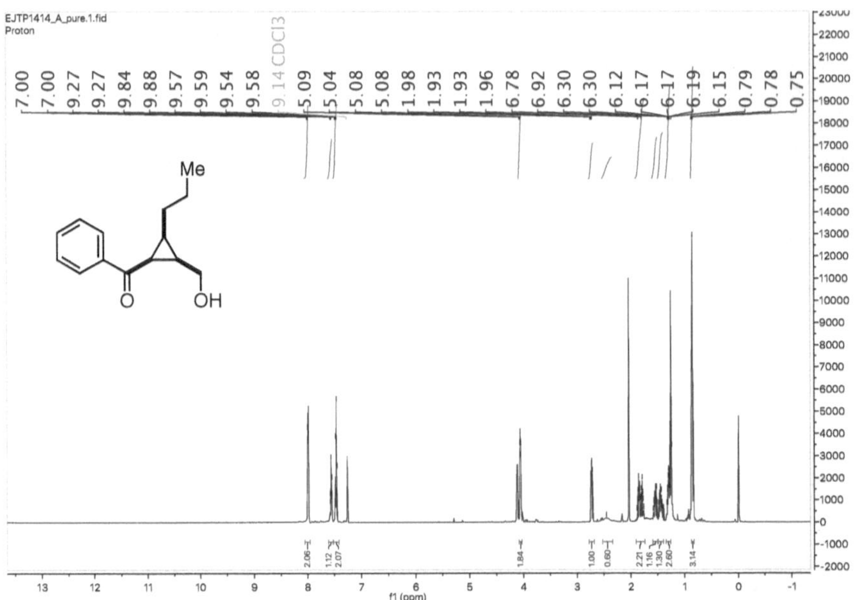

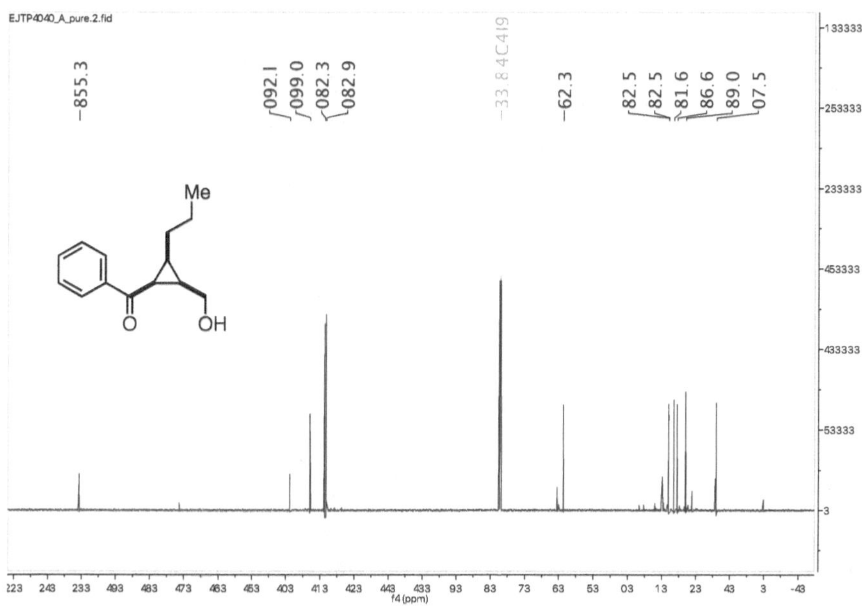

EJTP4040_A_pure.2.fid
Me
O
OH
855.3
092.1
099.0
082.3
082.9
62.3
82.5
82.5
81.6
86.6
89.0
07.5
f4 (ppm)

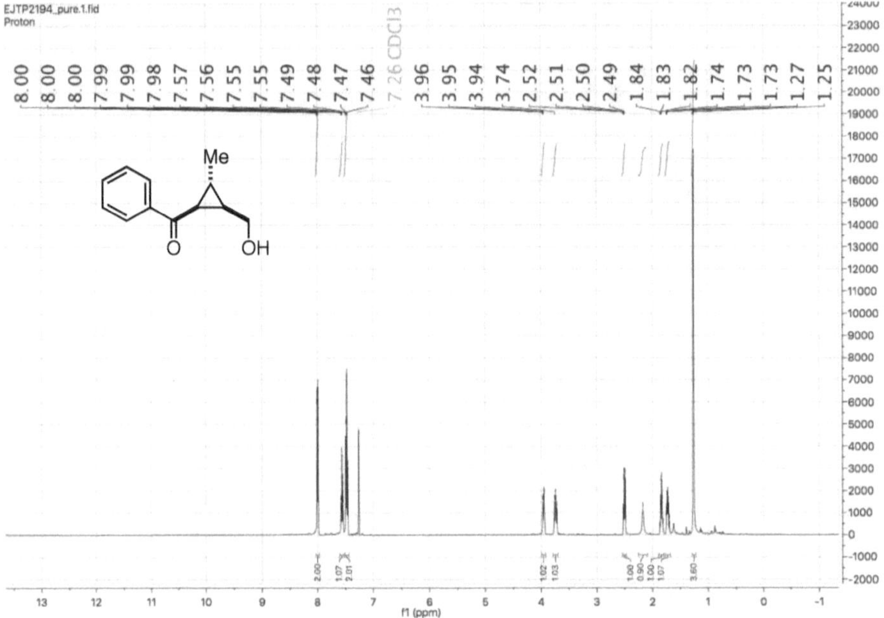

EJTP2194_pure.1.fid
Proton
7.26 CDCl3
Me
OH

EJTP2194_pure.2.fid
Carbon 13

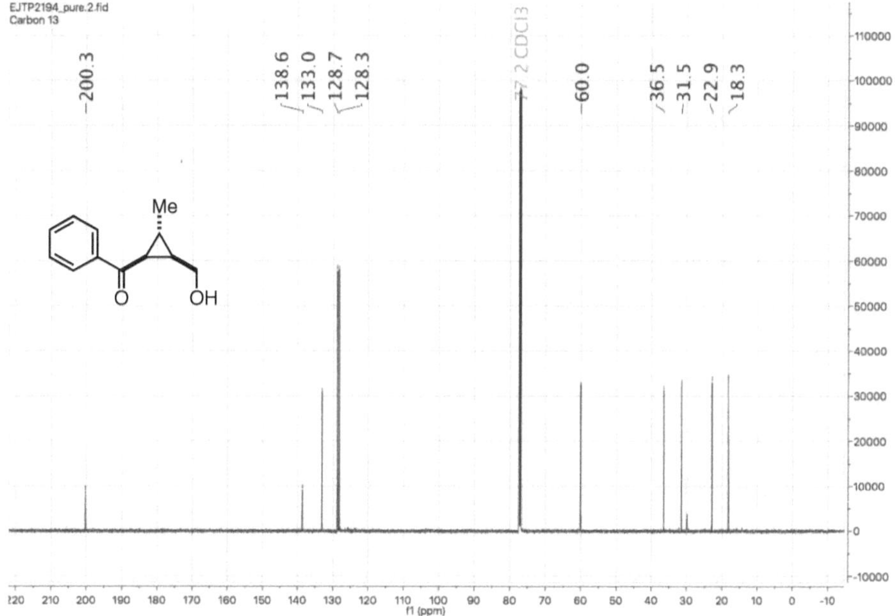

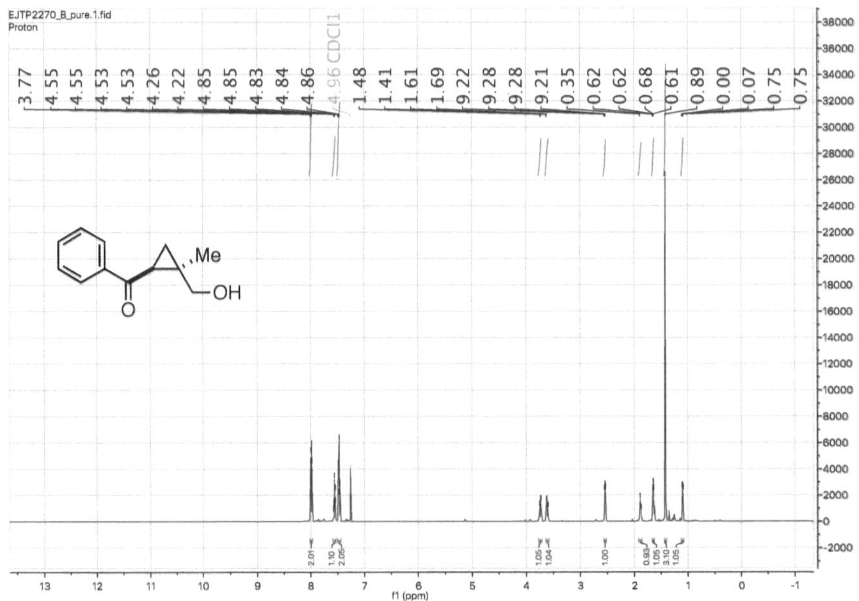

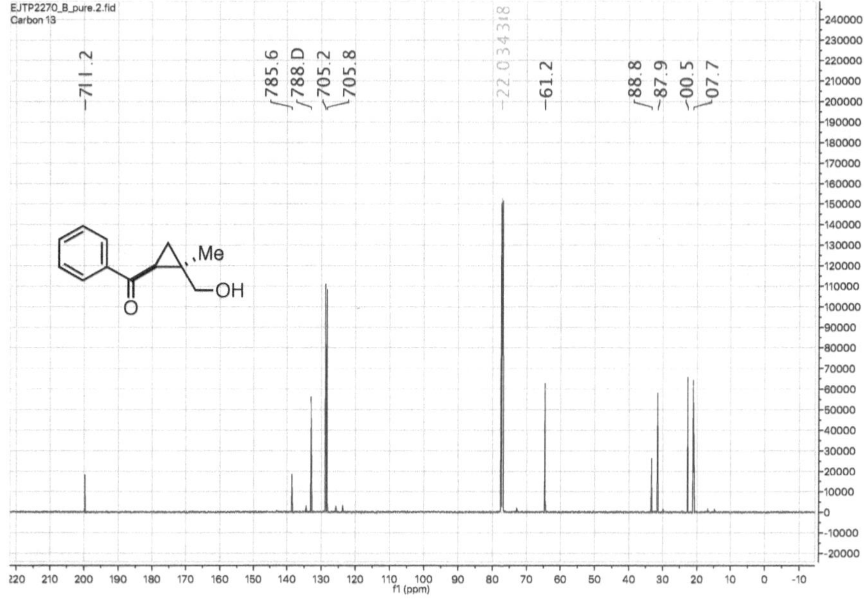

EJTP2270_B_pure.2.fid
Carbon 13
711.2
785.6
788.D
705.2
705.8
22.034318
61.2
88.8
87.9
00.5
07.7
Me
OH
O
240000
230000
220000
210000
200000
190000
180000
170000
160000
150000
140000
130000
120000
110000
100000
90000
80000
70000
60000
50000
40000
30000
20000
10000
0
-10000
-20000
220 210 200 190 180 170 160 150 140 130 120 110 100 90 80 70 60 50 40 30 20 10 0 -10
f1 (ppm)

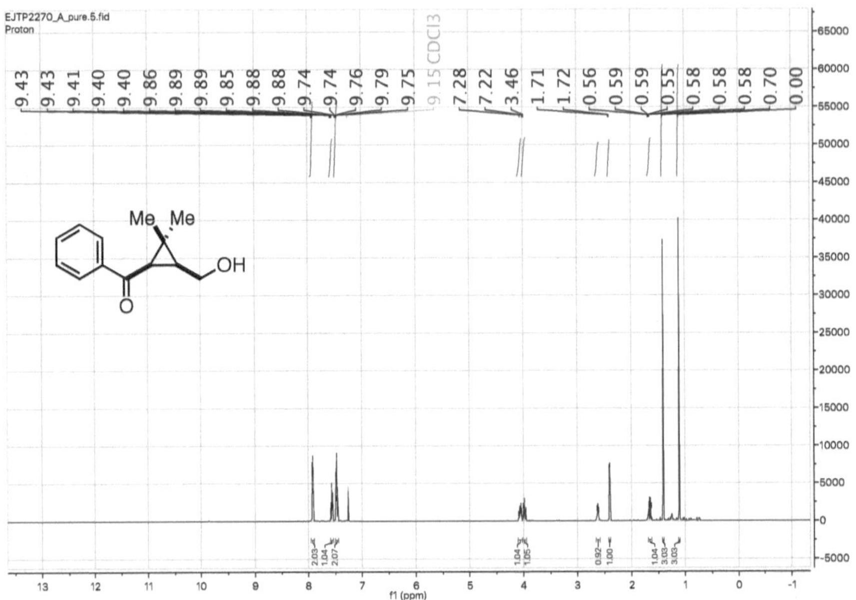

EJTP2270_A_pure.5.fid
Proton
9.15 CDCl3
9.43
9.43
9.41
9.40
9.40
9.86
9.89
9.89
9.85
9.88
9.88
9.74
9.74
9.76
9.79
9.75
7.28
7.22
3.46
1.71
1.72
0.56
0.59
0.59
0.55
0.58
0.58
0.58
0.70
0.00
Me Me
OH
O
f1 (ppm)

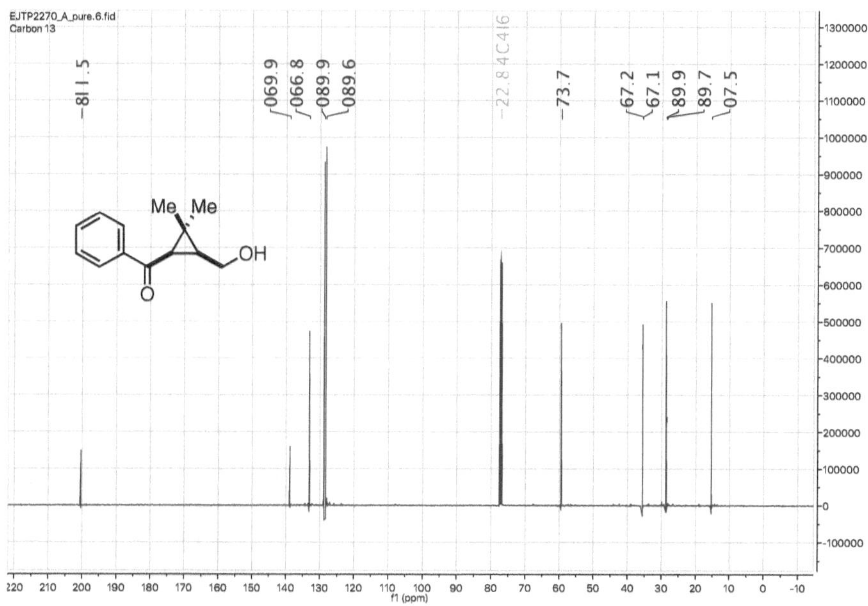

EJTP2270_A_pure.6.fid
Carbon 13

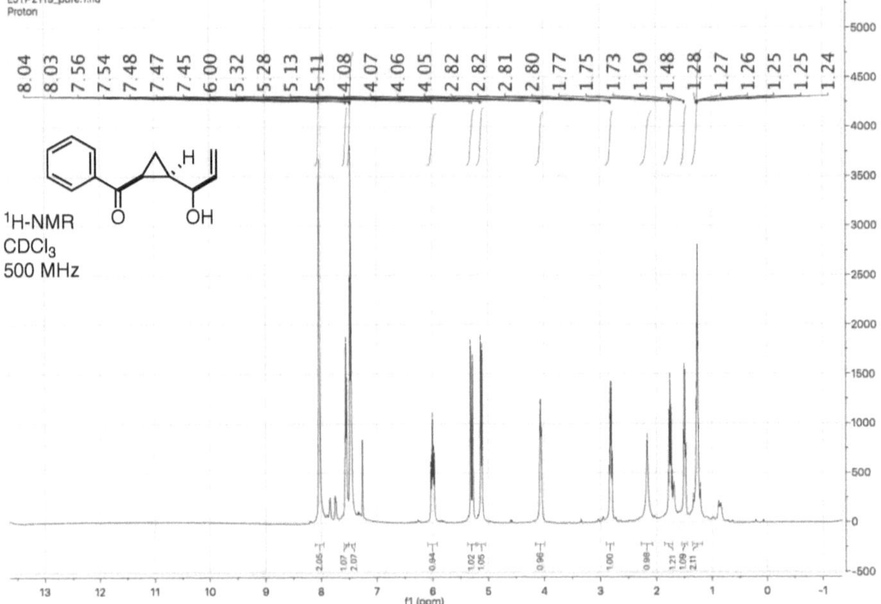

EJTP2119_pure.1.fid
Proton
1H-NMR
CDCl3
500 MHz

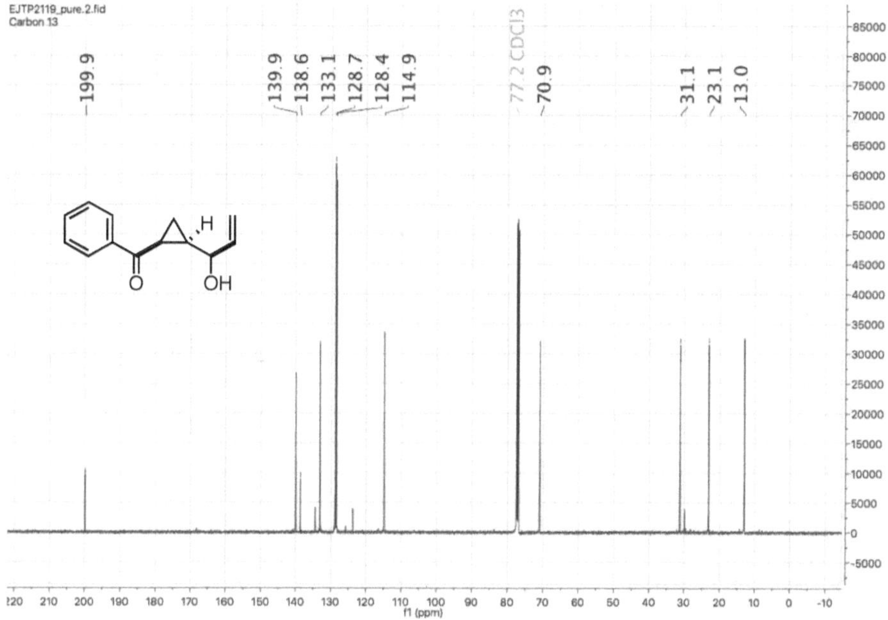

EJTP2119_pure.2.fid
Carbon 13
199.9
139.9
138.6
133.1
128.7
128.4
114.9
77.2 CDCl3
70.9
31.1
23.1
13.0
f1 (ppm)

EJTP2213_A_pure.1.fid
Proton

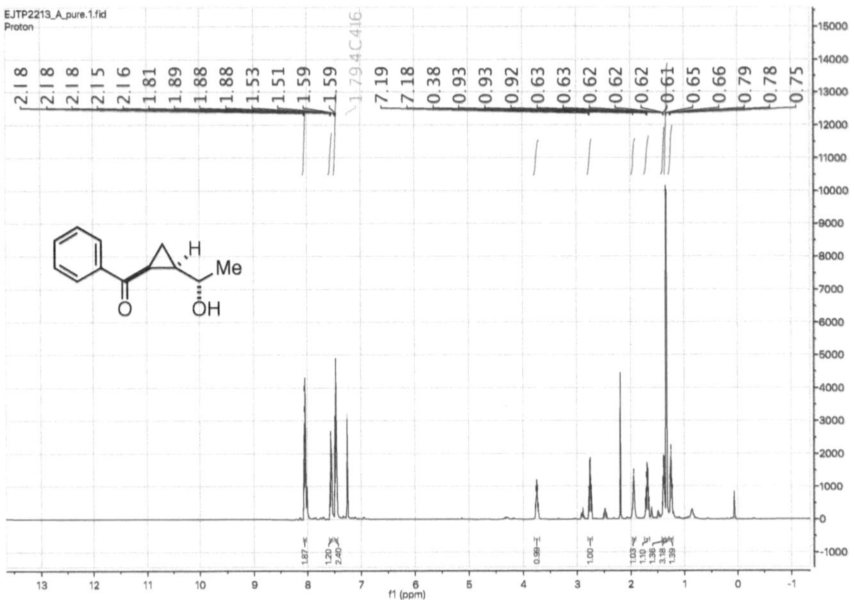

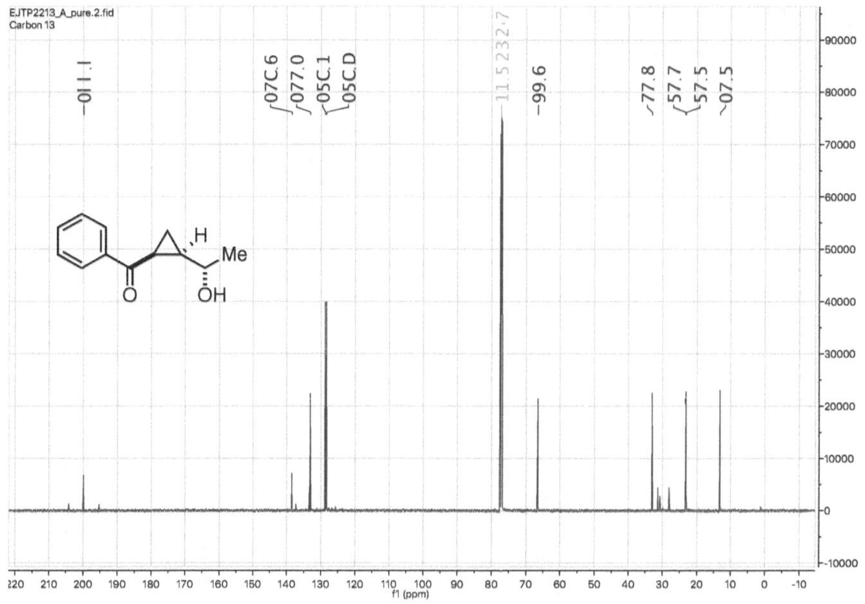

EJTP2213_A_pure.2.fid
Carbon 13
199.0
137.6
137.0
130.1
130.0
77.5
77.2
76.9
66.6
33.8
27.7
27.5
20.5
Me
H
OH
O
f1 (ppm)

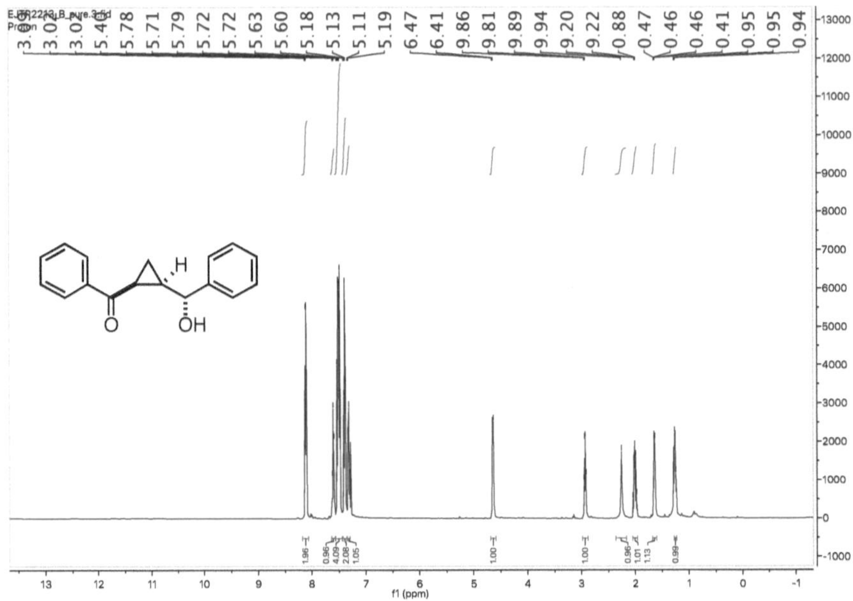

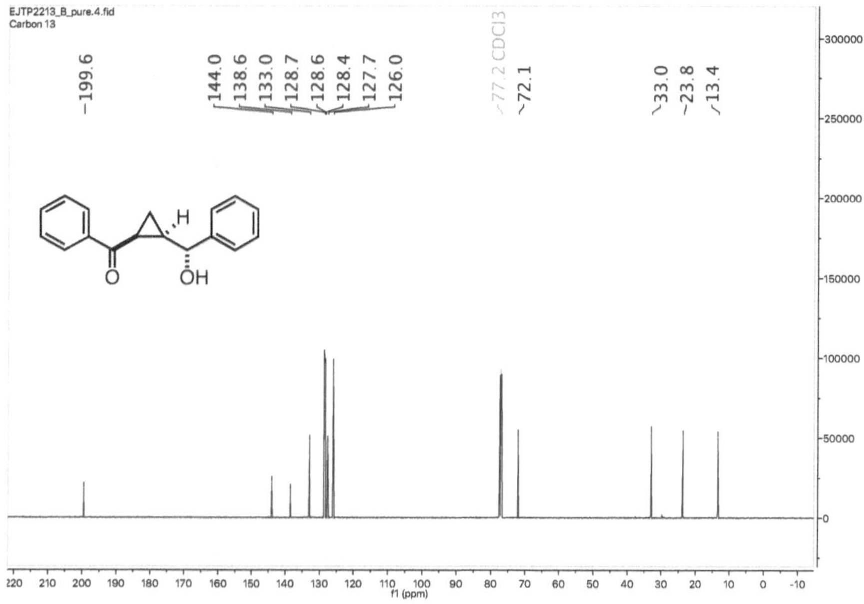

EJTP2213_B_pure.4.fid
Carbon 13
199.6
144.0
138.6
133.0
128.7
128.6
128.4
127.7
126.0
77.2 CDCl3
72.1
33.0
23.8
13.4

EJTP1458_pure.1.fid
Proton

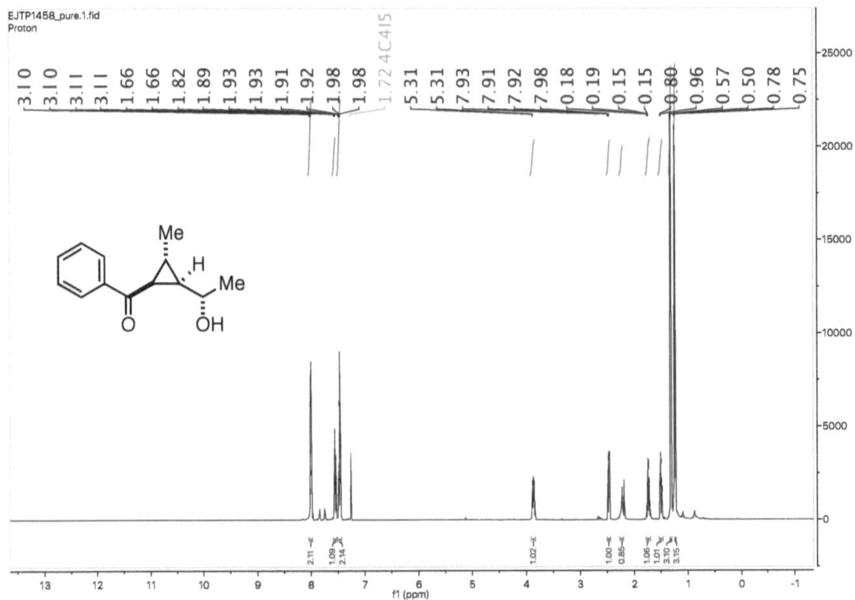

EJTP1458_pure.2.fid
Carbon 13
200.2
138.6
133.0
128.7
128.7
128.7
128.2
77.2 CDCl3
66.0
42.2
32.0
23.2
23.1
18.2
Me
H
Me
OH
O

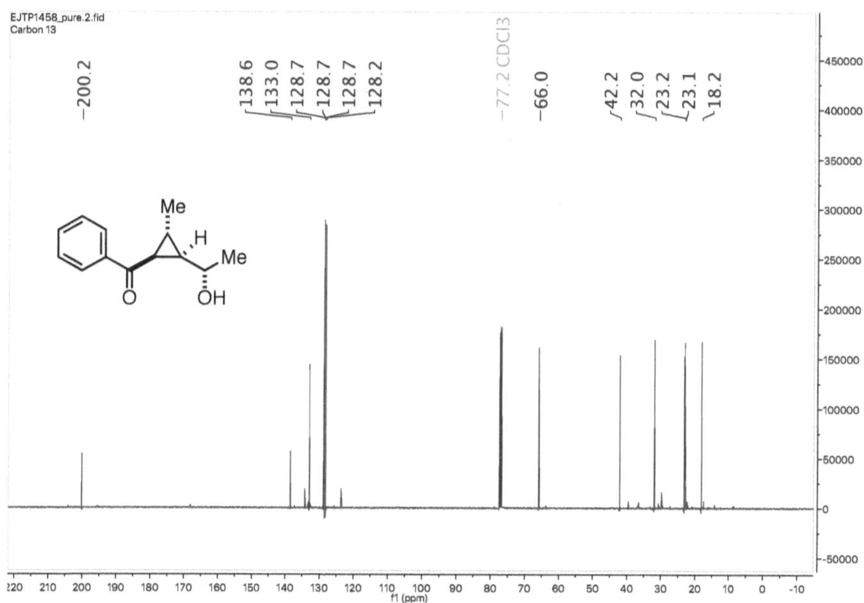

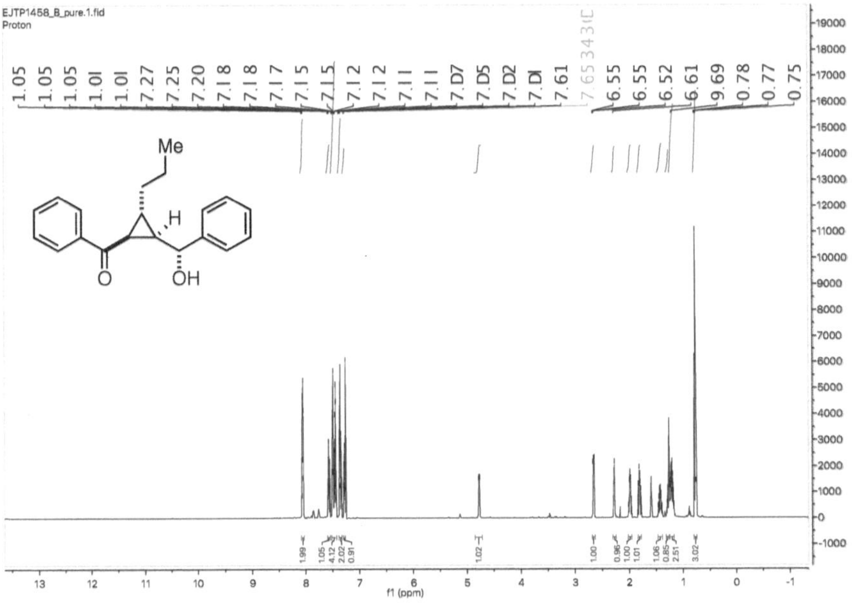

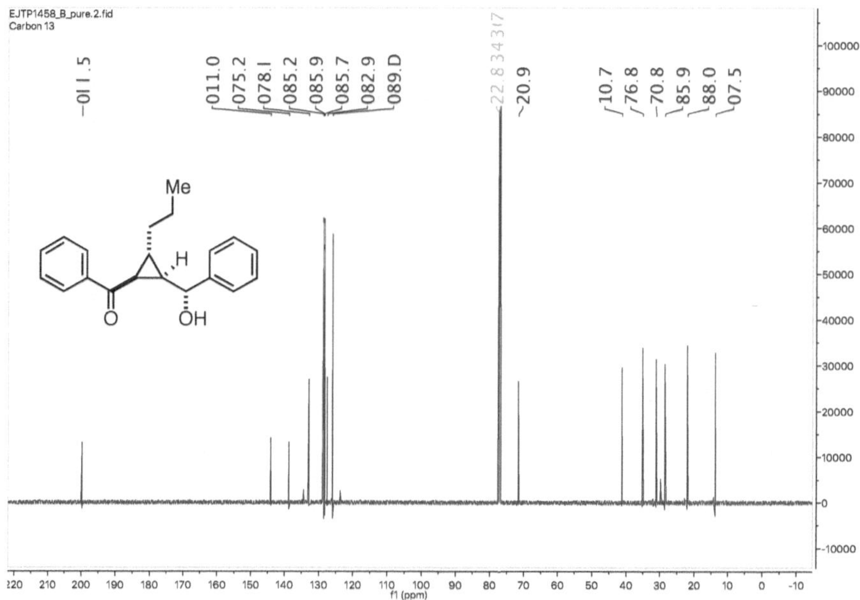

EJTP1458_B_pure.2.fid
Carbon 13
Me
H
OH

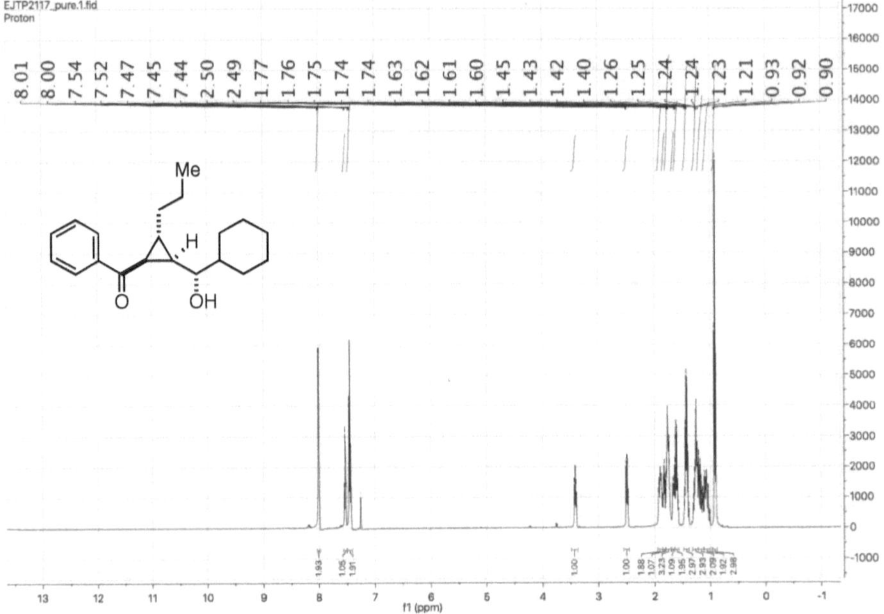

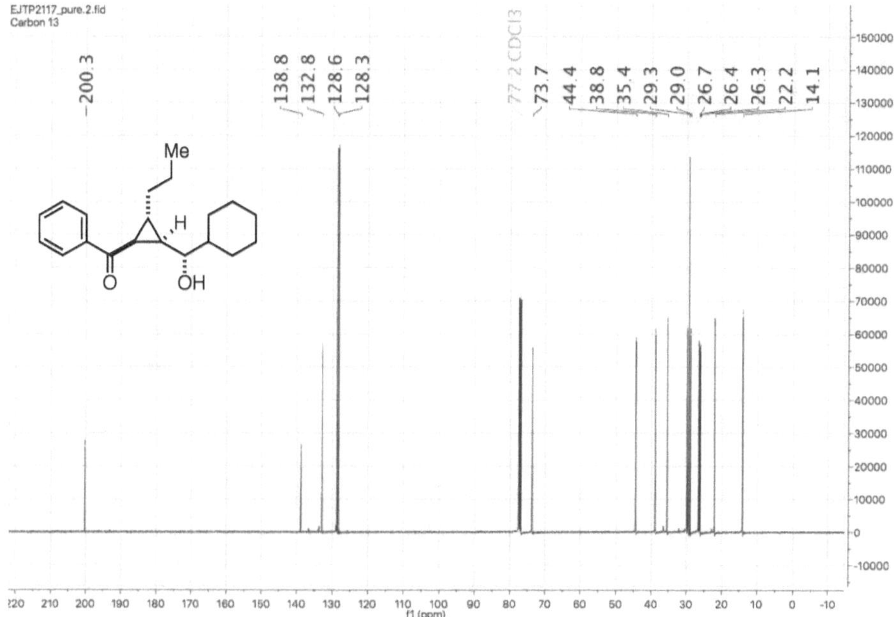

EJTP2117_pure.2.fid
Carbon 13
200.3
138.8
132.8
128.6
128.3
77.2 CDCl3
73.7
44.4
38.8
35.4
29.3
29.0
26.7
26.4
26.3
22.2
14.1
Me
H
OH
O

EJTP2318_pure.1.fid

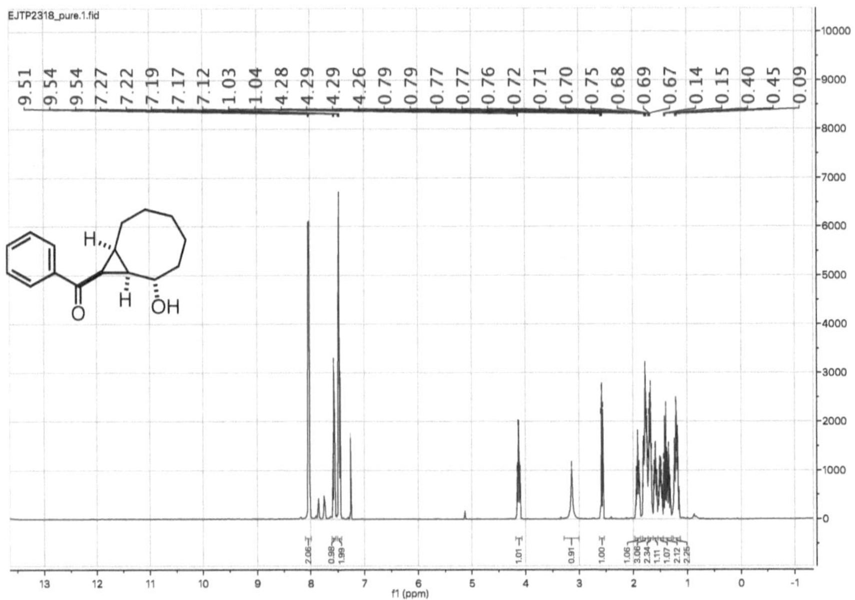

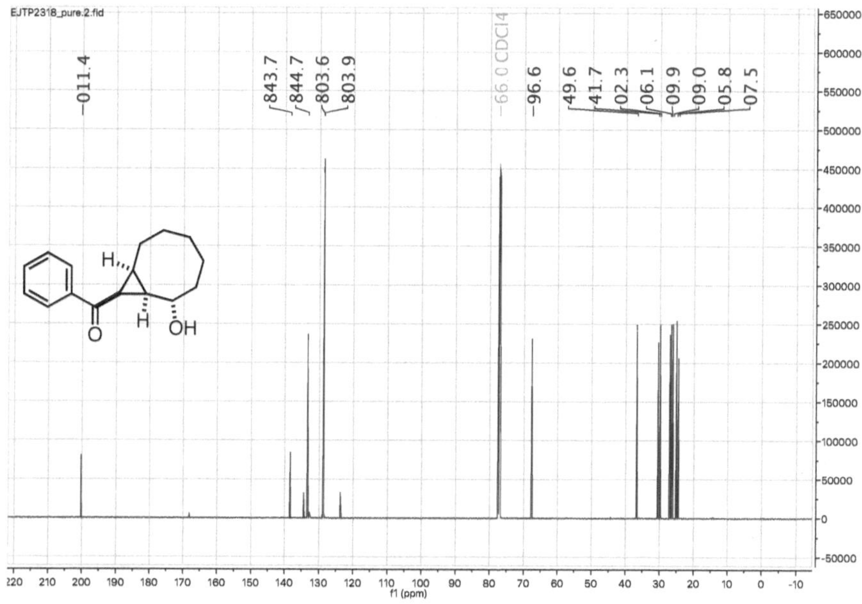

EJTP2318_pure.2.fid
011.4
843.7
844.7
803.6
803.9
66.0 CDCl4
96.6
49.6
41.7
02.3
06.1
09.9
09.0
05.8
07.5
f1 (ppm)

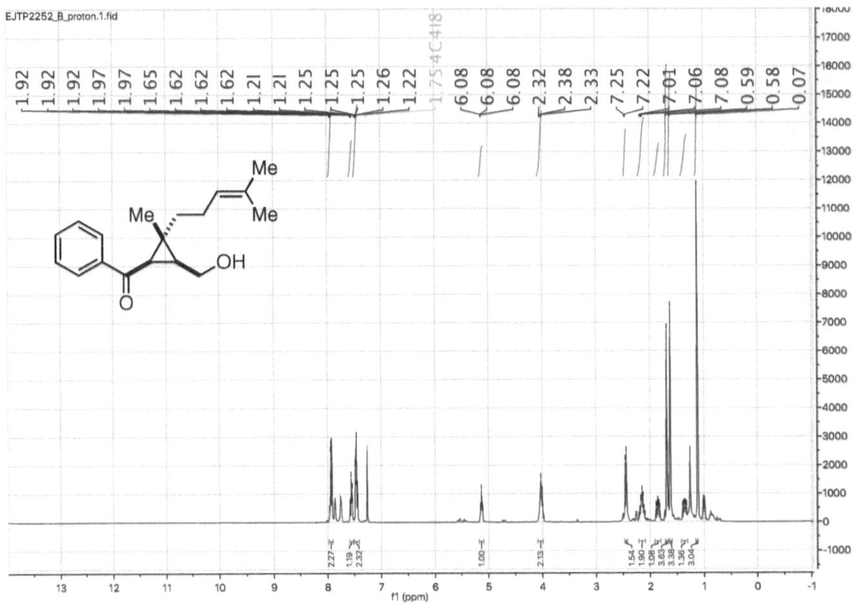

EJTP2252_B_proton.1.fid
Me
Me
Me
OH
O

EJTP2252_B_pure.7.fid
Carbon 13

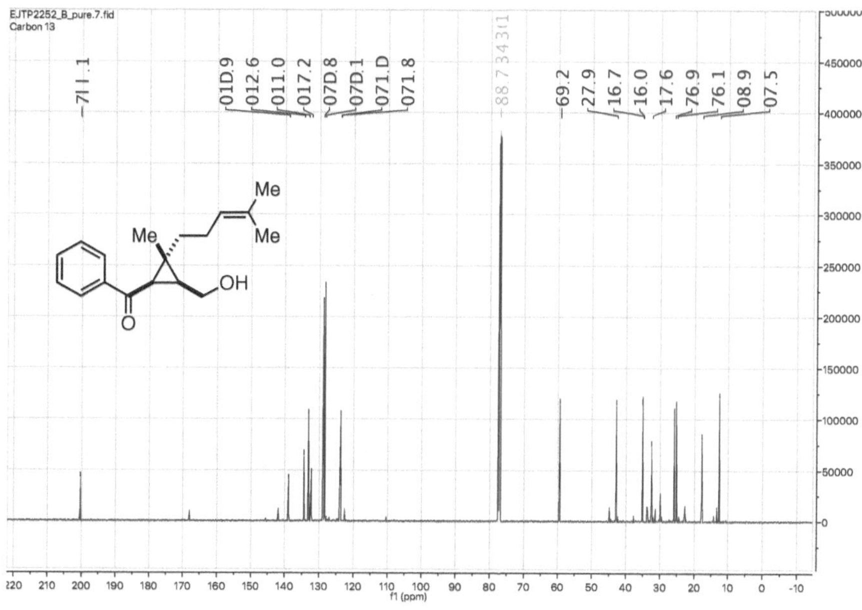

Me
Me
Me
Me
OH
O

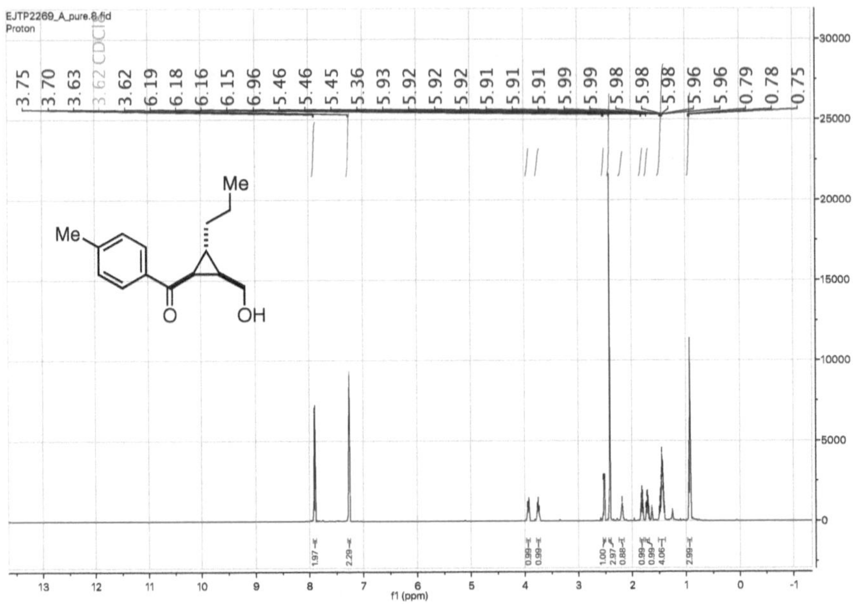
Me
Me
OH
O
3.75
3.70
3.63
3.62 CDCl
3.62
6.19
6.18
6.16
6.15
6.96
5.46
5.46
5.45
5.36
5.93
5.92
5.92
5.92
5.91
5.91
5.91
5.99
5.99
5.98
5.98
5.98
5.96
5.96
0.79
0.78
0.75
1.97
2.29
0.99
0.99
1.00
2.97
0.88
0.99
0.99
4.06
2.99
f1 (ppm)

EJTP2269_A_pure.9.fid
Carbon 13

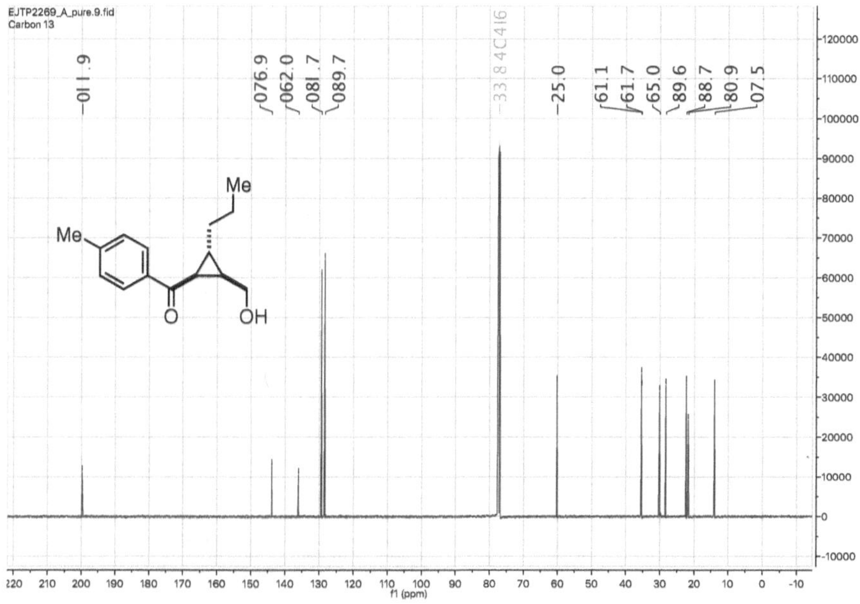

Me
Me
OH
O

EJTP2269_B_pure.5.fid
Proton

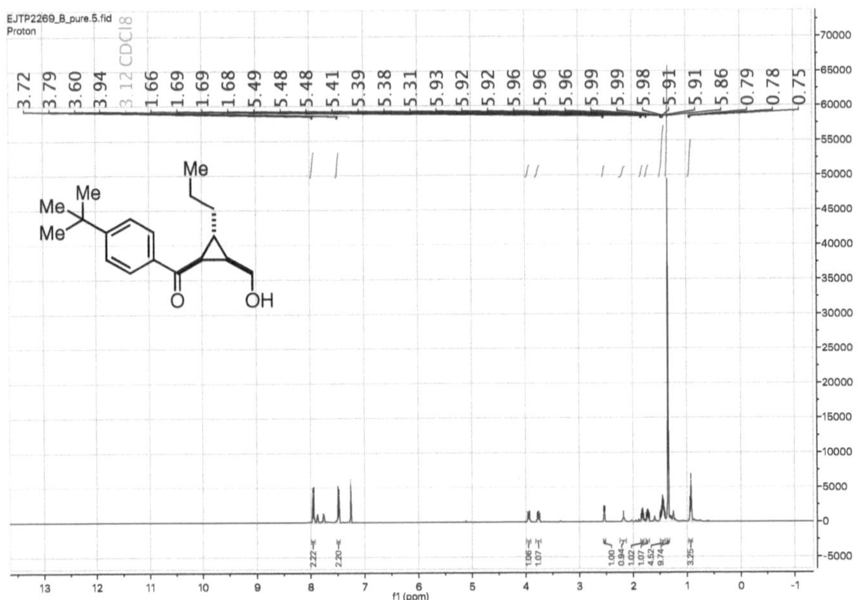

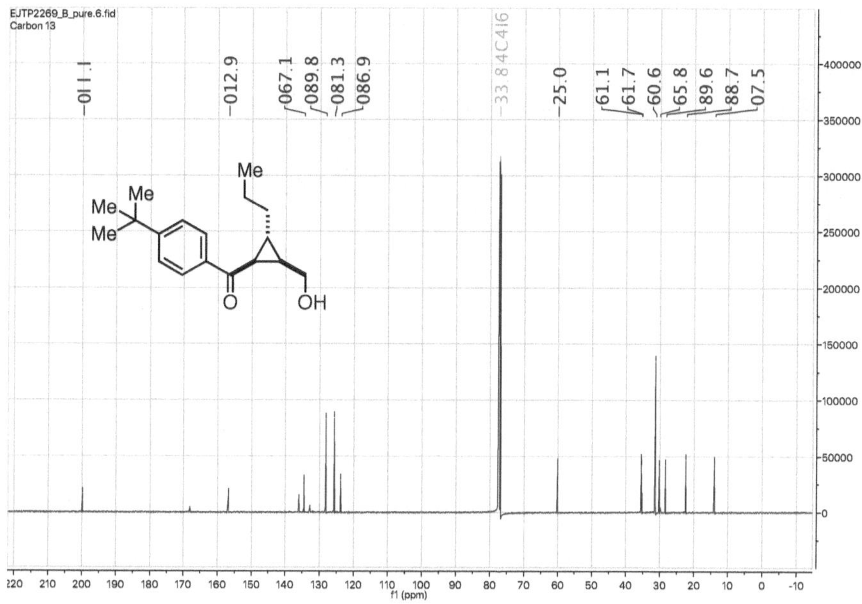

EJTP2269_B_pure.6.fid
Carbon 13
Me
Me Me
Me
OH
O
f1 (ppm)

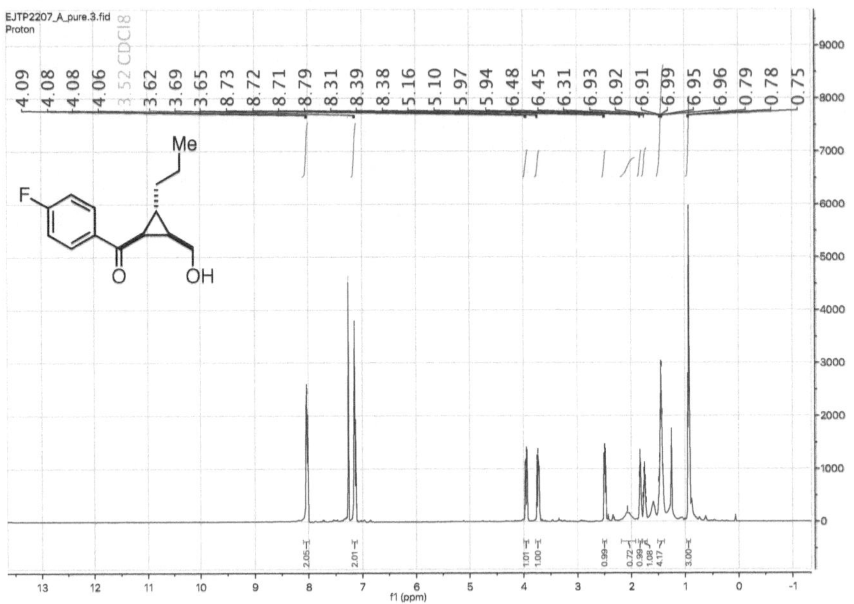

EJTP2207_A_pure.4.fid
Carbon 13

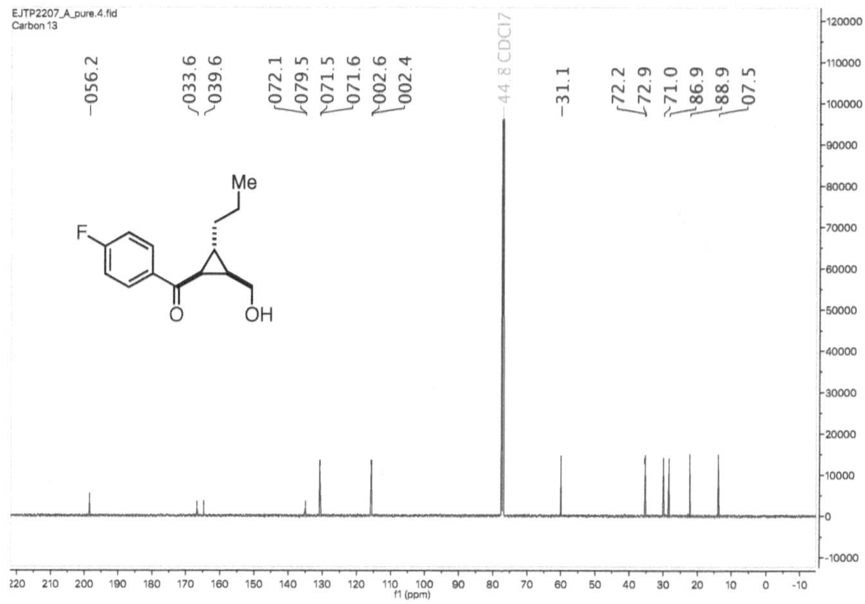

EJTP2104_Fluorine.2.fid
-104.90
-104.92
-104.93
-104.95
-104.96
-104.98
-104.99
Me
F
O
OH
f1 (ppm)

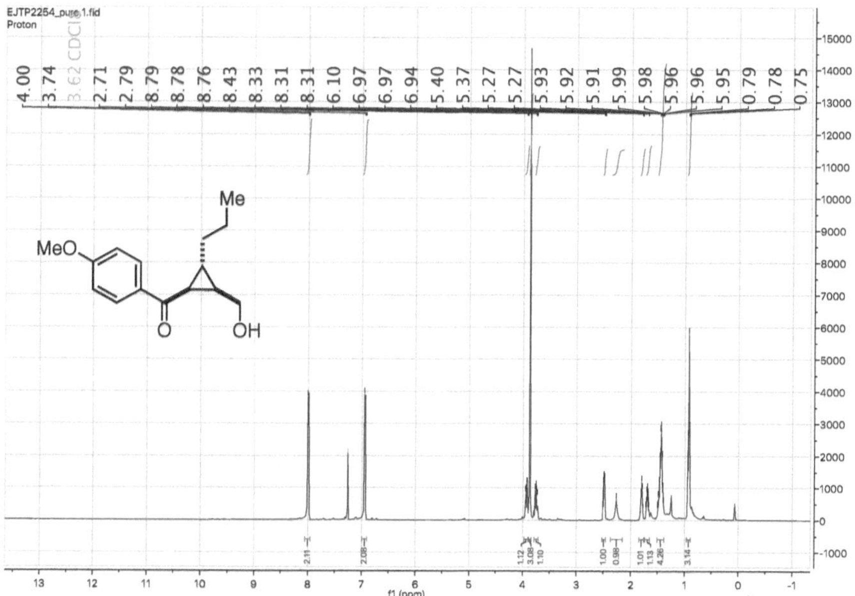

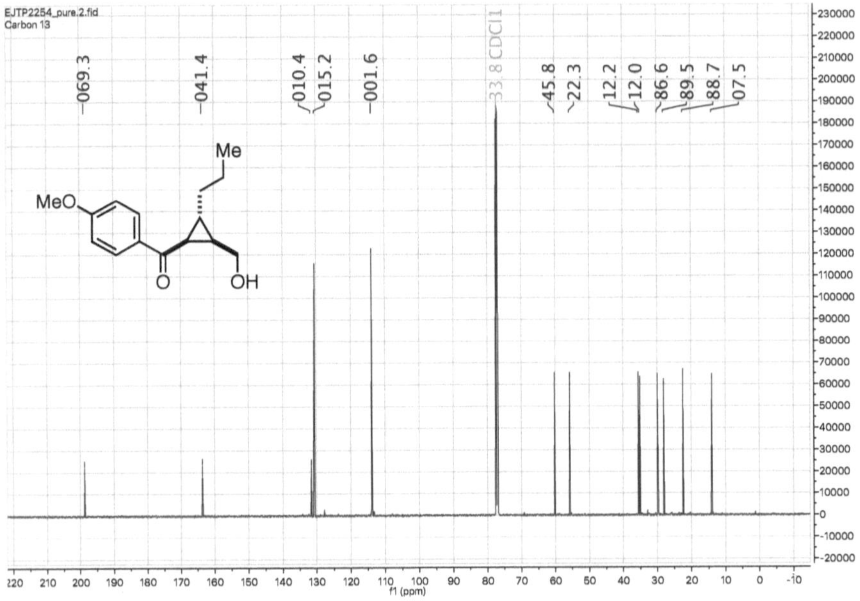

EJTP2254_pure.2.fid
Carbon 13
069.3
041.4
010.4
015.2
001.6
33.8 CDCl
45.8
22.3
12.2
12.0
86.6
89.5
88.7
07.5
MeO
Me
OH
O
f1 (ppm)

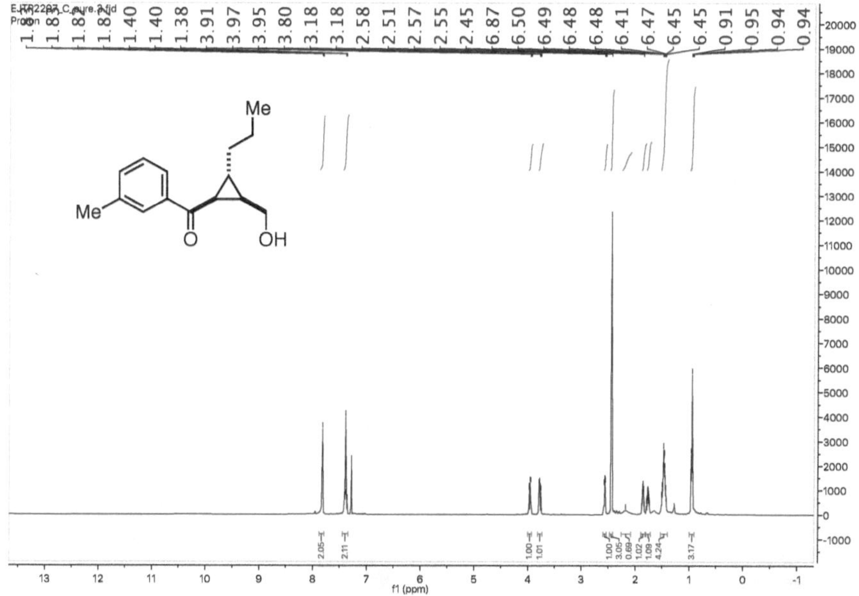

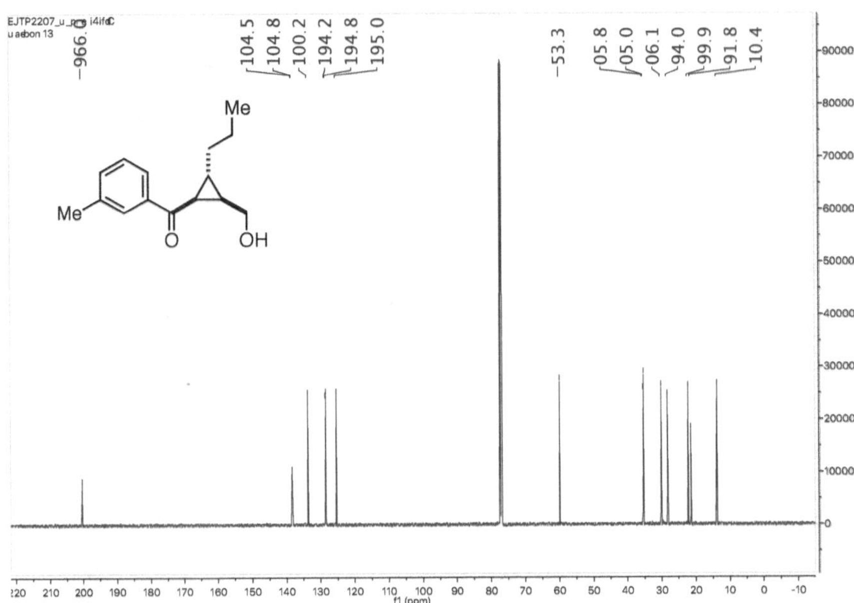

EJTP2207_u_p i4ifc
u aebon 13
966.
104.5
104.8
100.2
194.2
194.8
195.0
53.3
05.8
05.0
06.1
94.0
99.9
91.8
10.4
Me
Me
OH
O
f1 (ppm)

EJTP2207_B_pure.1.fid
Proton

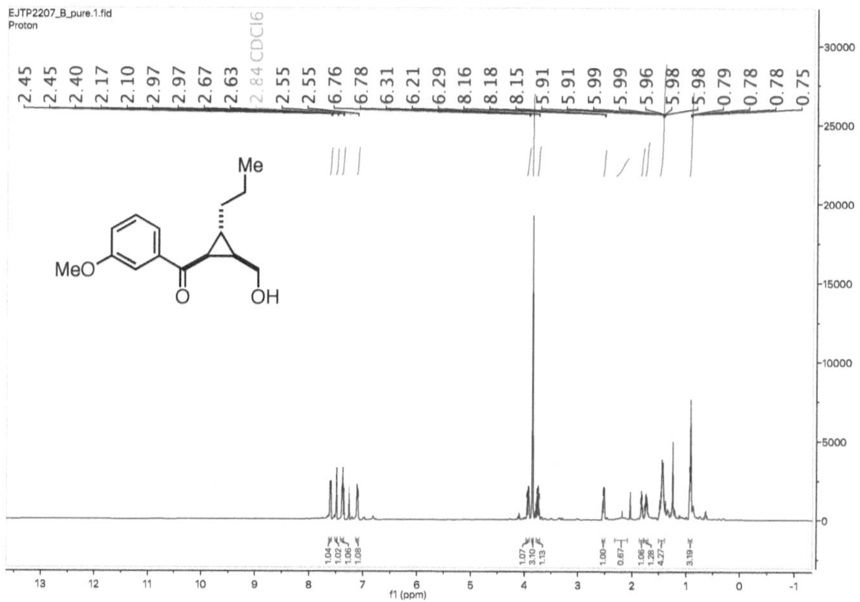

EJTP2207_B_pure.2.fid
Carbon 13

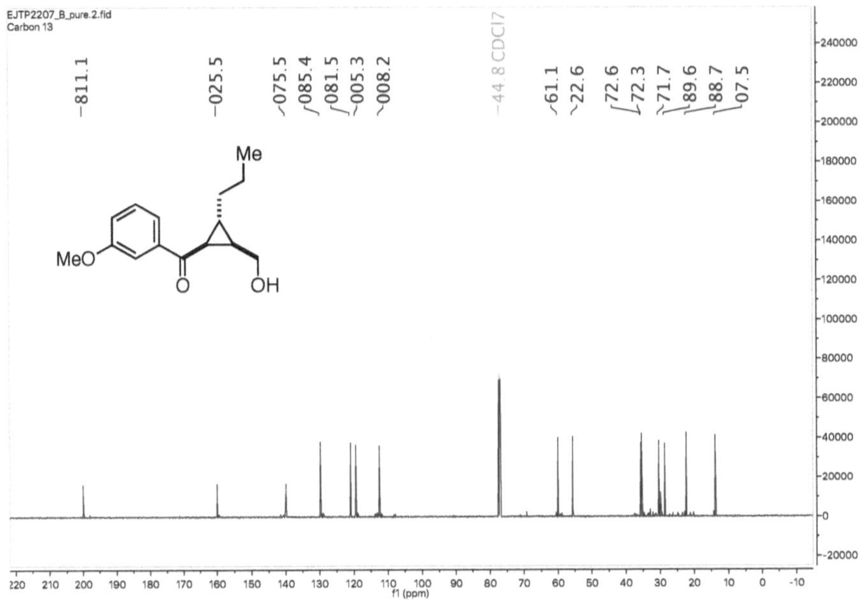

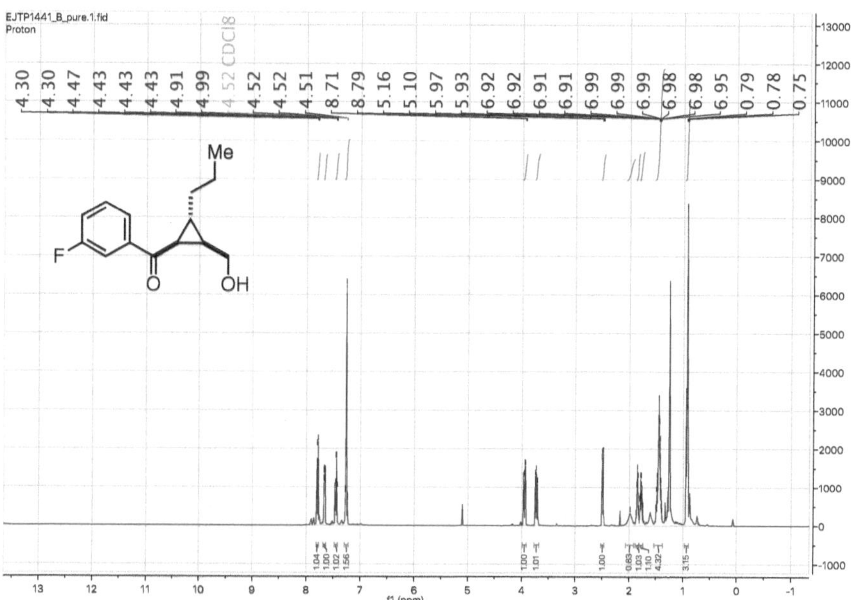

EJTP1441_B_pure.1.fid
Proton
4.52 CDCl3
Me
F
O
OH

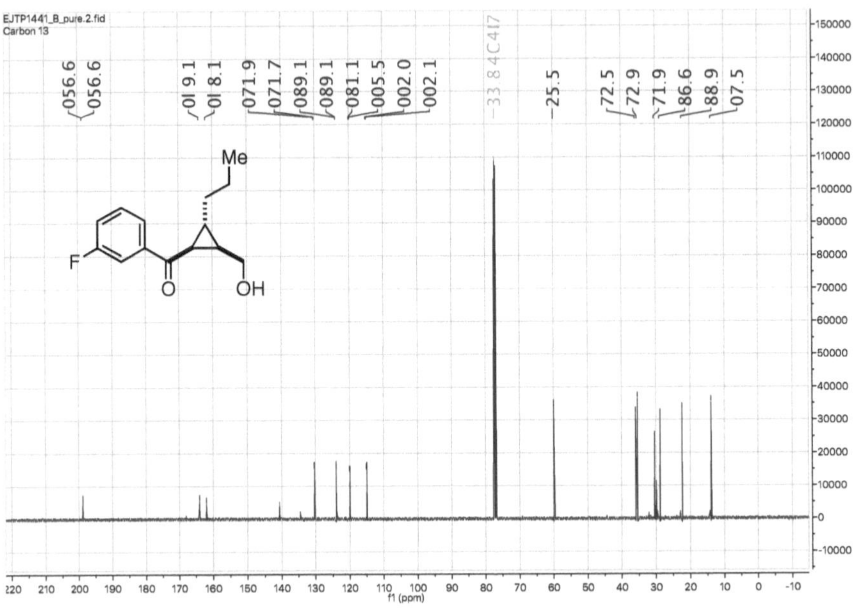

EJTP1441_B_pure.2.fid
Carbon 13
Me
F
O
OH
196.6
196.6
169.1
168.1
129.1
127.7
118.9
118.9
116.1
115.5
112.0
112.1
33.84C4l7
25.5
22.5
22.9
21.9
16.6
18.9
07.5
150000
140000
130000
120000
110000
100000
90000
80000
70000
60000
50000
40000
30000
20000
10000
0
-10000
220 210 200 190 180 170 160 150 140 130 120 110 100 90 80 70 60 50 40 30 20 10 0 -10
f1 (ppm)

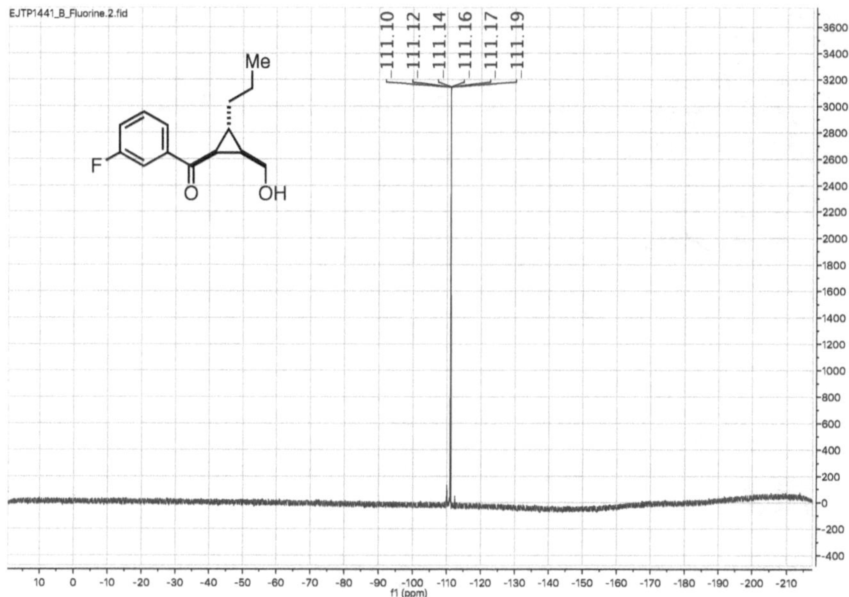

EJTP1441_B_Fluorine.2.fid
Me
F
O
OH
111.10
111.12
111.14
111.16
111.17
111.19

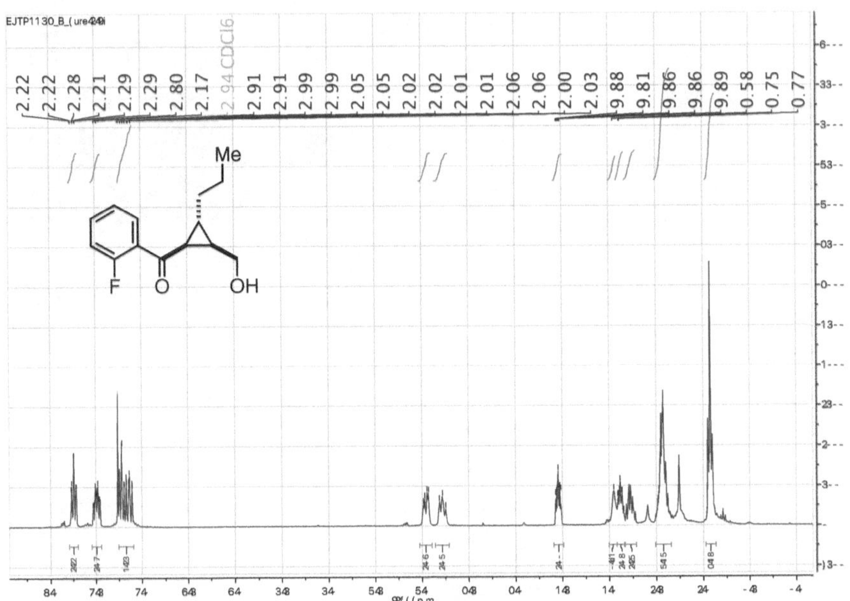

EJTP2269_D_pure.6.fid
Carbon 13
Me
F
O
OH
f1 (ppm)

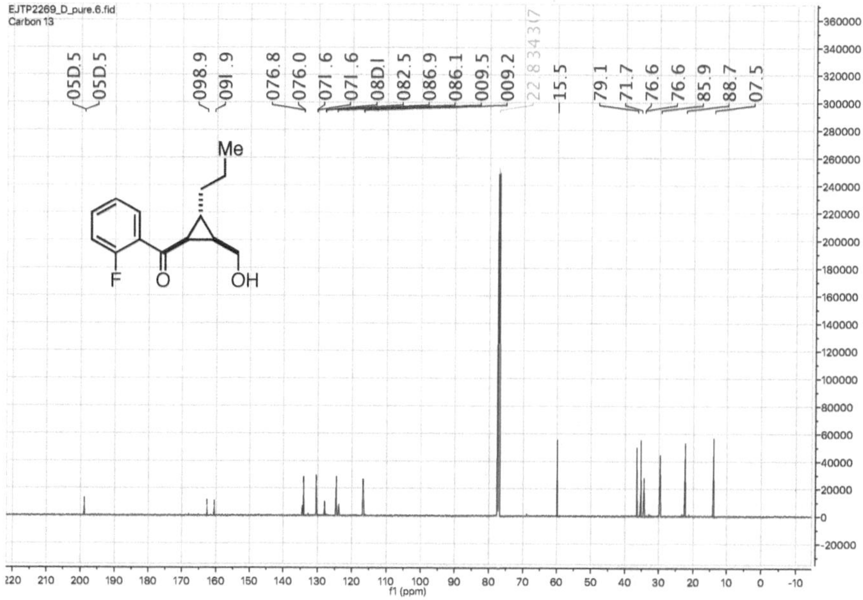

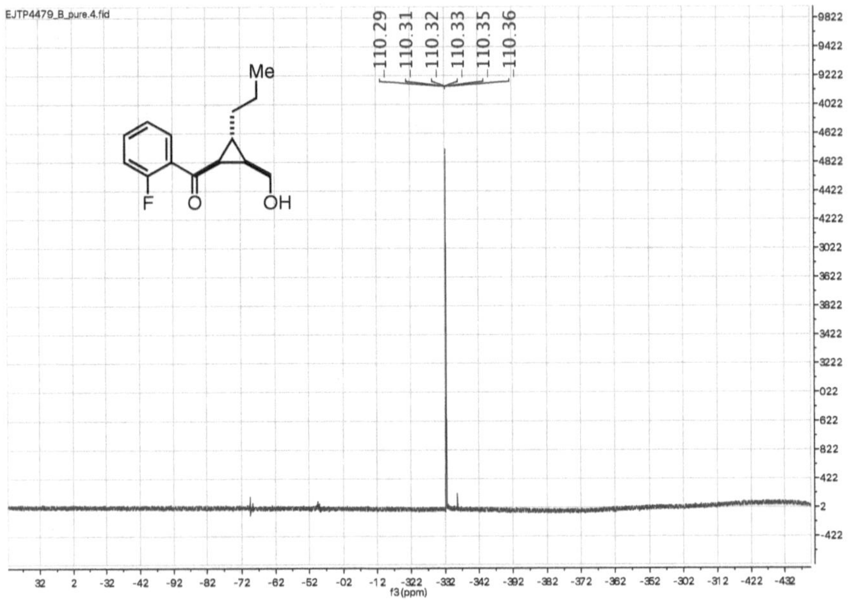

EJTP4479_B_pure.4.fid
Me
F
O
OH
110.29
110.31
110.32
110.33
110.35
110.36

EJTP1441_A_pure.1.fid
Proton

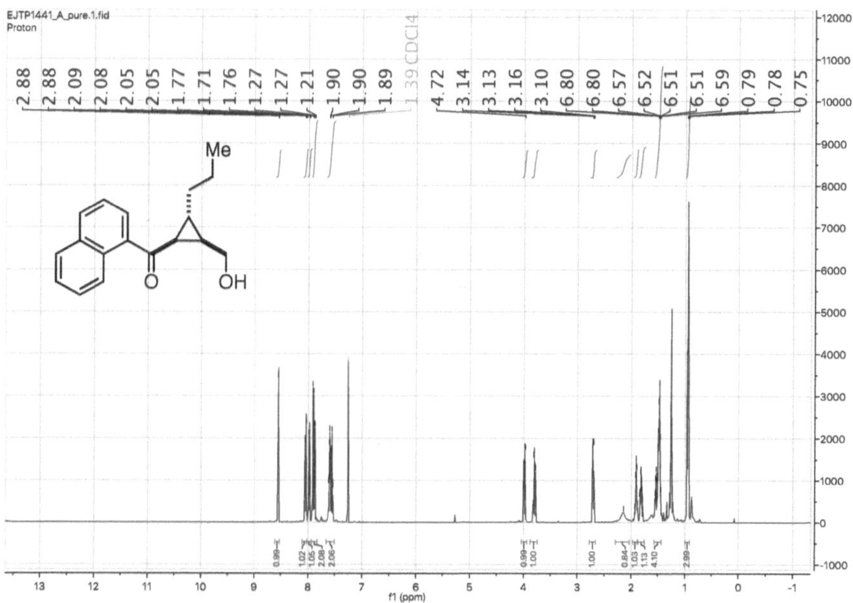

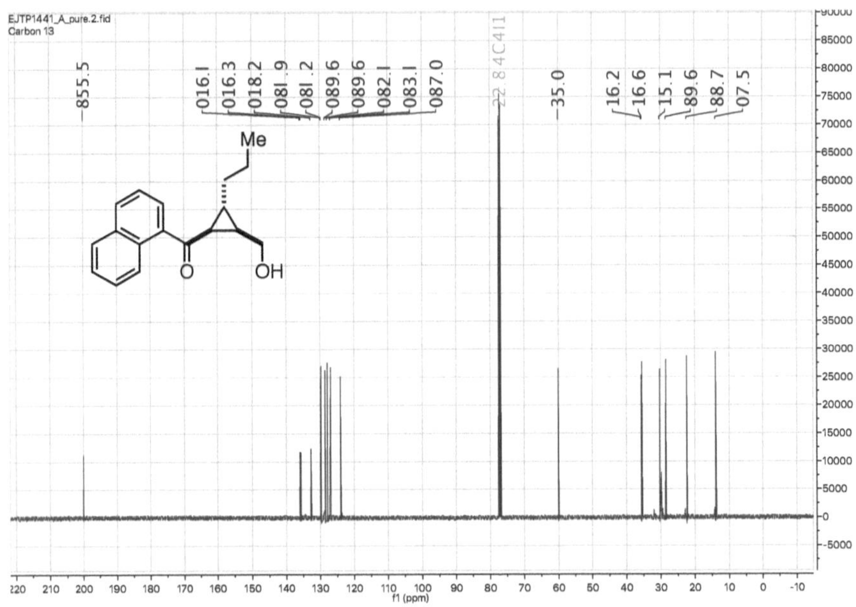

EJTP1441_A_pure.2.fid
Carbon 13
855.5
016.1
016.3
018.2
081.9
081.2
089.6
089.6
082.1
083.1
087.0
22.84C411
35.0
16.2
16.6
15.1
89.6
88.7
07.5
Me
OH
90000
85000
80000
75000
70000
65000
60000
55000
50000
45000
40000
35000
30000
25000
20000
15000
10000
5000
0
-5000
220 210 200 190 180 170 160 150 140 130 120 110 100 90 80 70 60 50 40 30 20 10 0 -10
f1 (ppm)

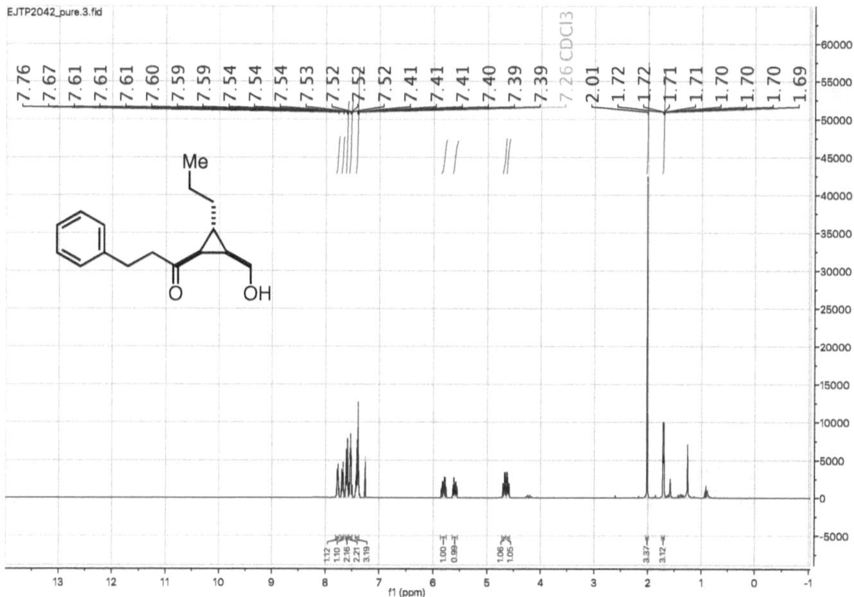

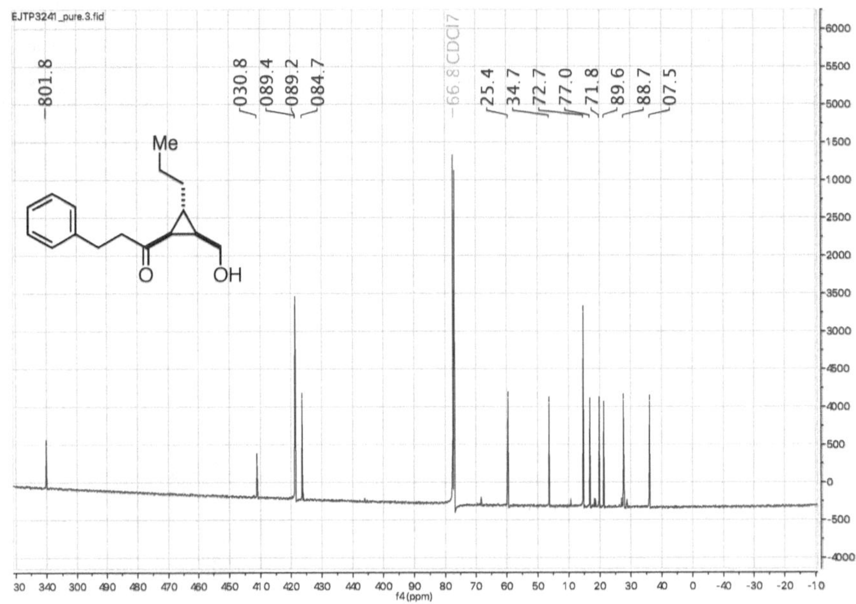

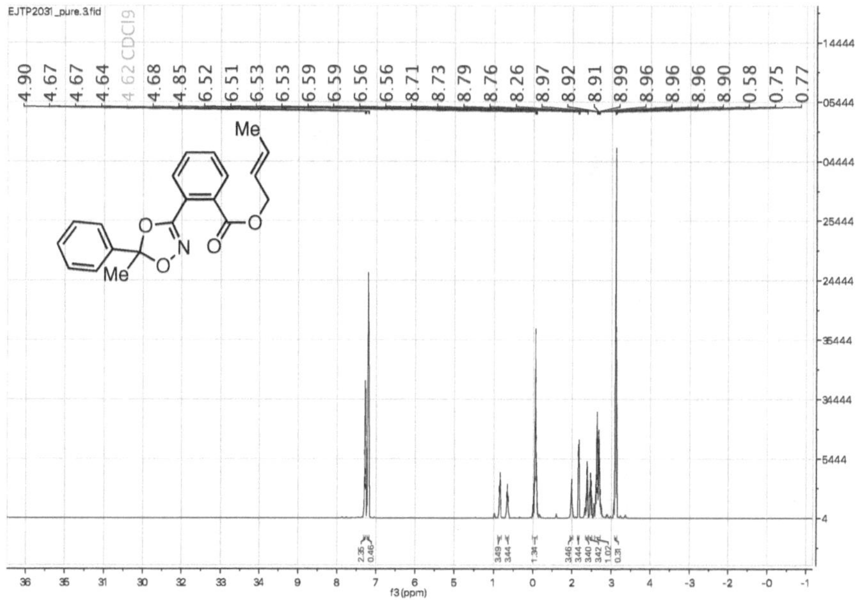

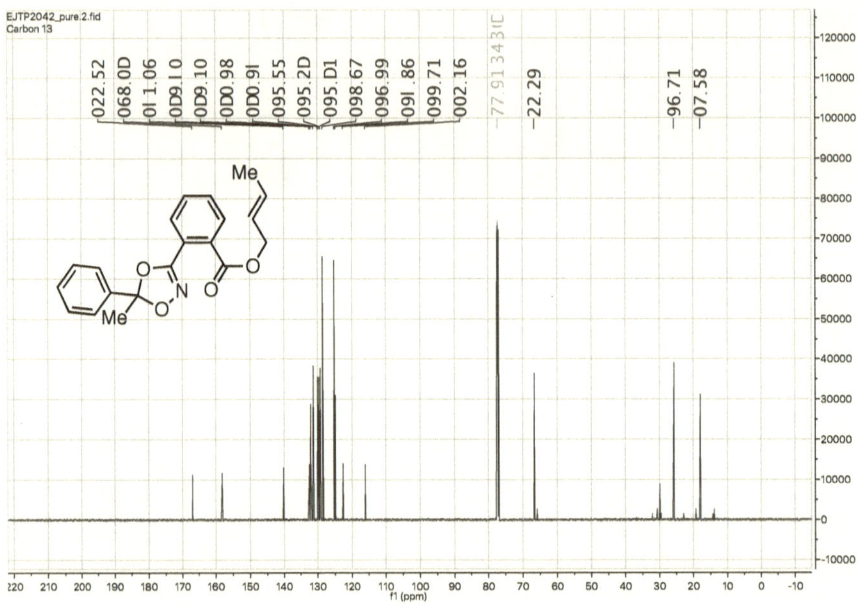

EJTP2042_pure.2.fid
Carbon 13
Me
Me
f1 (ppm)

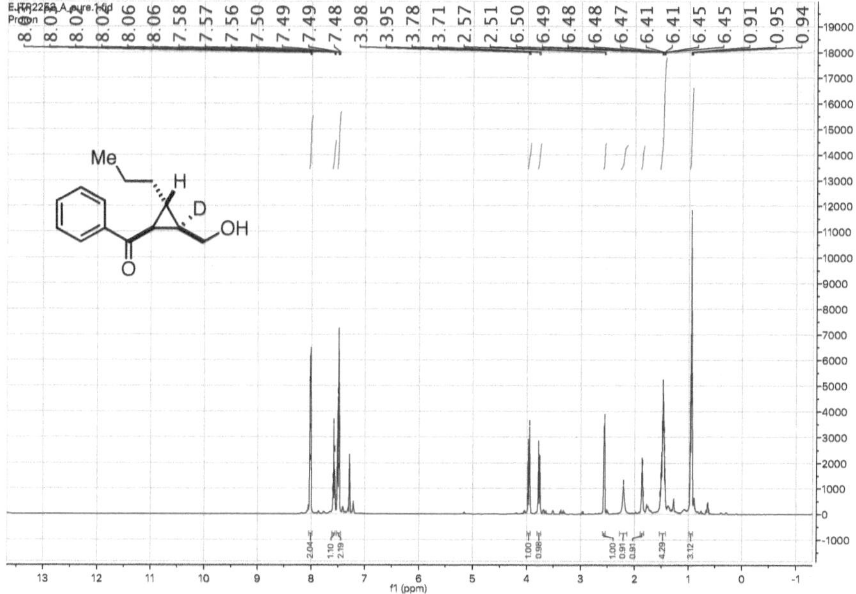

E.TR225A_A_pure.1.fid
Proton
Me
H
D
OH
O
8.07 8.07 8.06 8.06 7.58 7.57 7.56 7.50 7.49 7.49 7.48 3.98 3.95 3.78 3.71 2.57 2.51 1.50 1.49 1.48 1.48 1.47 1.41 1.41 1.45 1.45 0.91 0.95 0.94
f1 (ppm)

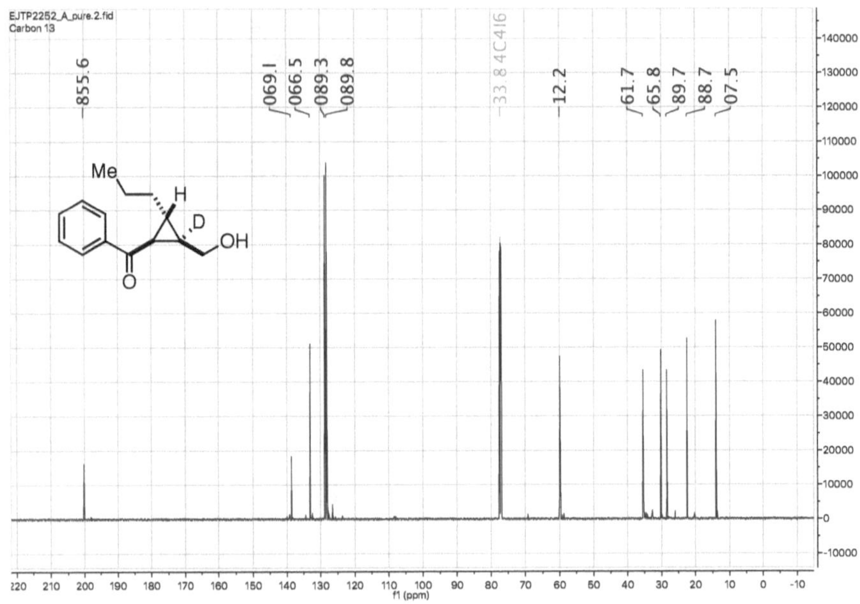

EJTP2252_A_pure.2.fid
Carbon 13
855.6
069.1
066.5
089.3
089.8
33.84C416
12.2
61.7
65.8
89.7
88.7
07.5
Me
H
D
OH
O
f1 (ppm)

A2.8 References

(1)   Still, W. C.; Kahn, M.; Mitra, A. J. Org. Chem. 1978, 43, 2923–2925.

(2)   Gassman, P.G.; Sowa, J.R. 1,2,3,4-Tetraalkyl-5-perfluoroalkyl-cyclopentadiene, di-(perfluoroalkyl)-trialkylcyclopentadiene and transition metal complexes thereof, U.S. Patent 5,245,064, Sep. 14, 1993.

(3)   **3a-3j**: Piou, T.; Rovis, T. *J. Am. Chem. Soc.,* **2014**, *136*, 11292.

(4)   **3k:** Duchemin, C.; Cramer, N. *Org. Chem. Front.,* **2019**, *6*, 209.

(5)   Kim, J.D.; Lee, M.H.; Han, G.; Park, H.; Zee, O.P.; Jung, Y.H. *Tetrahedron,* **2001**, *57*, 8257.

(6)   **2j:** Tasukawa, T.; Miyamura, H.; Kobayashi, S. *J. Am. Chem. Soc.,* **2012**, *134*, 16963.

(7)   **2k:** Wonk, K.C.; Ng, E.; Wong, W.-T.; Chiu, P. *Chem. Eur. J.,* **2016**, *22*, 3709.

(8)   **Cyclooctenol:** Li, J.; Jia, S.; Chen, P. R. *Nature Chemical Biology,* **2014**, *10*, 1003.

(9)   **2a-d₁:** Fox, R.J; Lalic, G; Bergman, R.G. *J. Am. Chem. Soc.,* **2007**, *129*, 14144.

(10)   **8a:** Park, S. R.; Kim, C.; Kim, D.; Thrimurtulu, N.; Yeom, H.-S.; Jun, J.; Shin, S.;Rhee, Y.H. *Org. Lett.,* **2013**, *15*, 1166.

(11)   **9a:** Motokuni, K; Takeuchi, D.; Osakada, K.; *Polym. Chem.,* **2015**, *6*, 1248.

(12)   Blanc, E.; Schwarzenbach, D.; Flack, H. D. *J. Appl. Cryst.* **24** (1991), 1035-1041.

(13)   Clark. R. C.; Reid, J. S. *Acta Cryst.* **A51** (1995), 887-897.

(14)   Version 1.171.38.46 (2015). Rigaku Oxford Diffraction.

(15)   Sheldrick, G. M. *Acta Cryst.* **A71** (2015), 3-8.

(16)   Sheldrick, G. M. *Acta Cryst.* **C71** (2015), 3-8.

(17)   Dolomanov, O. V.; Bourhis, L. J.; Gildea, R. J.; Howard, J. A. K.; Puschmann, H. *J. Appl. Cryst.* **42** (2009), 339-341.

(18)   Spek, A. *Acta Cryst.* **D65** (2009), 148-155.

(19)   CrystalMaker Software Ltd, Oxford, England *(www.crystalmaker.com)*.

– Appendix C –

Supporting Information for Chapter Four and Five

In a flame-dried 1-dram vial equipped with a magnetic stir bar, **4-1a** (5 equiv.), [Cp*^{CF3}RhCl$_2$]$_2$ (1 equiv.), KOPiv (2 equiv.) were added and weighed in air. 2,2,2-trifluoroethanol (TFE) (0.2M) was added via micropipette and the vial was sealed with a teflon cap. The atmosphere was then replaced with ethylene gas and stirred at room temperature overnight. TFE was removed and the crude residue was purified by flash chromatography with silica eluting with Hexanes: Ethyl Acetate (9:1 to 4:1) giving the desired metal complex an orange oil. A crystal was grown by taking up **4-4** in DCM and layering pentane on top and letting the vial rest in the freezer (~-20 °C) overnight. The structure was proposed by the crystallographer and final data is still being worked up.

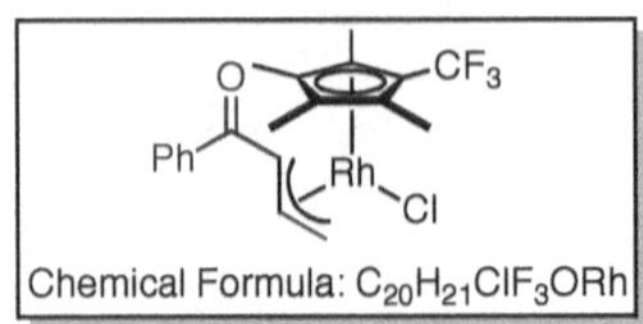

Yield: 83%

¹**H NMR** (400 MHz, Chloroform-*d*) δ 8.21 – 8.14 (m, 2H), 7.60 – 7.43 (m, 3H), 5.37 – 5.26 (m, 1H), 4.78 (d, *J* = 10.2 Hz, 1H), 4.35 (d, *J* = 7.3 Hz, 1H), 3.61 (d, *J* = 12.4 Hz, 1H), 2.00 (d, *J* = 1.3 Hz, 3H), 1.58 (s, 3H), 1.49 (d, *J* = 1.1 Hz, 3H), 1.28 (s, 3H).

¹⁹**F NMR** (282 MHz, CDCl₃) δ -53.36

LRMS m/z (ESI APCI) calculated for $C_{20}H_{21}ClF_3ORh$ [M+H] 473.0, found 473.0.

In a flame-dried 1-dram vial equipped with a magnetic stir bar, **5-5a** (2 equiv.), [Cp*MCl$_2$]$_2$ (0.5 equiv.), AgOAc (2 equiv.) were added and weighed in air. Methanol (0.2M) was added via micropipette and the vial was sealed with a teflon cap. The vial was then put in an aluminum heating block and stirred at 65 °C overnight. After letting the reaction cool, methanol was removed and the crude residue was purified by flash chromatography with silica eluting with DCM:MeOH (1% to 3% to 5%) giving the desired metal complex usually as an oil. Attempts at recrystallization are underway.

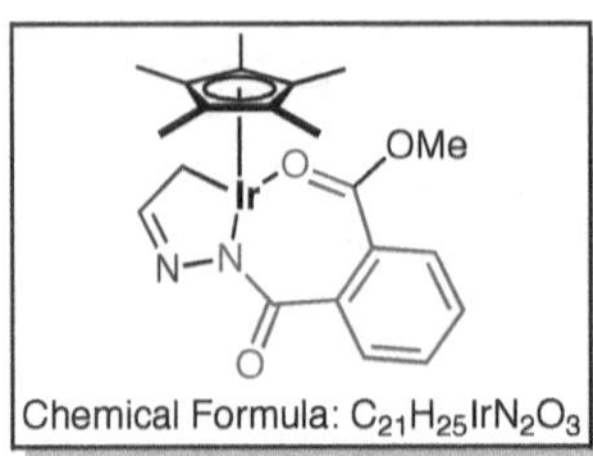

Yield: 61%

^{1}H NMR (500 MHz, Chloroform-*d*) δ 7.88 – 7.83 (m, 1H), 7.60 – 7.55 (m, 1H), 7.45 – 7.36 (m, 2H), 7.30 (q, *J* = 5.6 Hz, 1H), 3.82 (s, 3H), 2.46 (d, *J* = 5.7 Hz, 3H), 1.69 (s, 17H).

^{13}C NMR (126 MHz, Chloroform-*d*) δ 177.8, 169.7, 155.8, 133.4, 132.0, 130.4, 130.4, 129.7, 128.3, 85.2, 52.4, 18.2, 8.9

LRMS m/z (ESI APCI) calculated for C$_{21}$H$_{25}$N$_2$O$_3$Ir[M+H] 547.2, found 547.2.

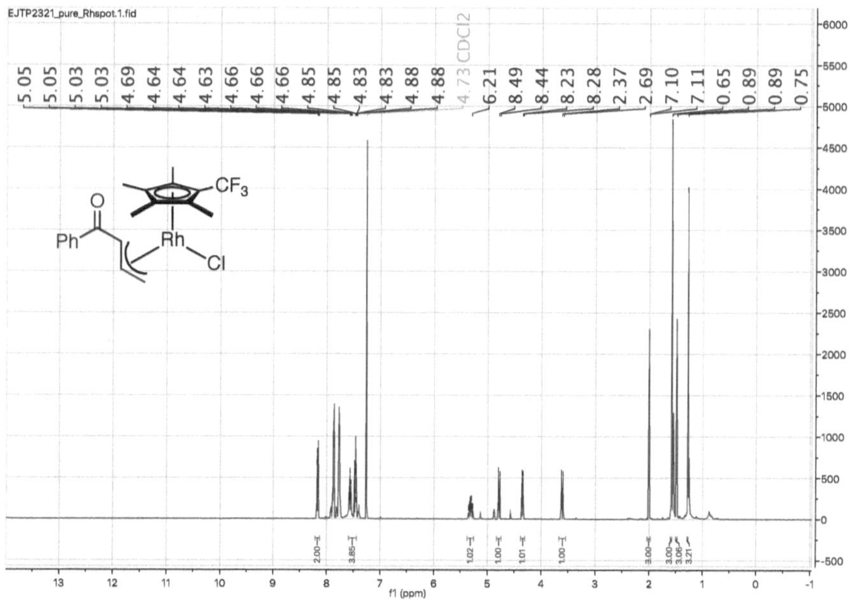

EJTP2321_pure_Rhspot.1.fid
5.05 5.05 5.03 5.03 4.69 4.64 4.64 4.63 4.66 4.66 4.66 4.85 4.85 4.83 4.83 4.88 4.88
4.73 CDCl2
6.21 8.49 8.44 8.23 8.28 2.37 2.69 7.10 7.11 0.65 0.89 0.89 0.75
Ph
CF3
Rh
Cl
f1 (ppm)

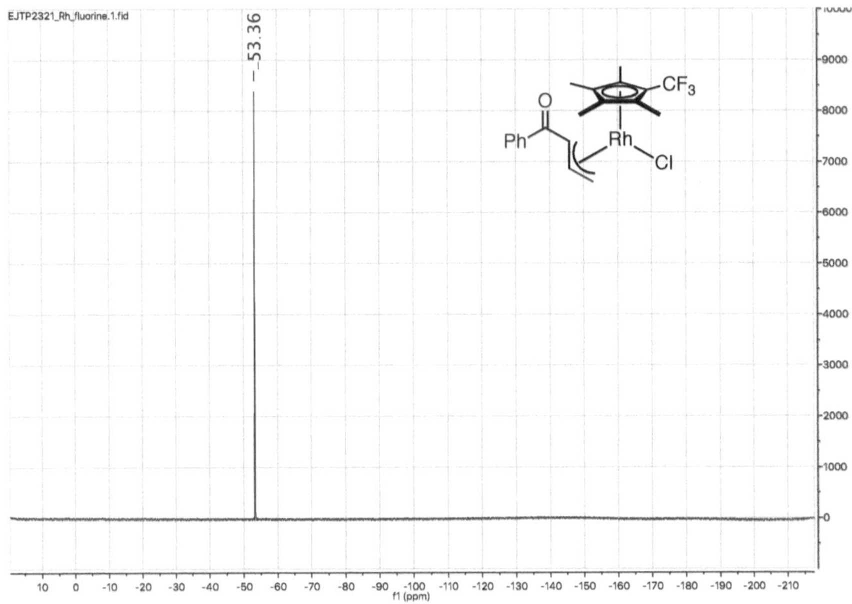

EJTP2321_Rh_fluorine.1.fid
−53.36
CF₃
Ph
O
Rh
Cl
f1 (ppm)

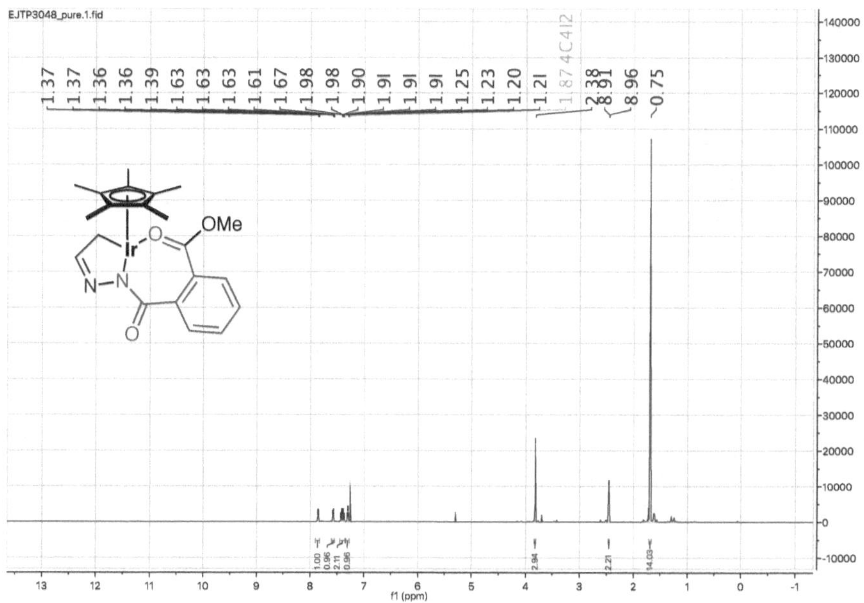

EJTP3048_pure.1.fid
OMe

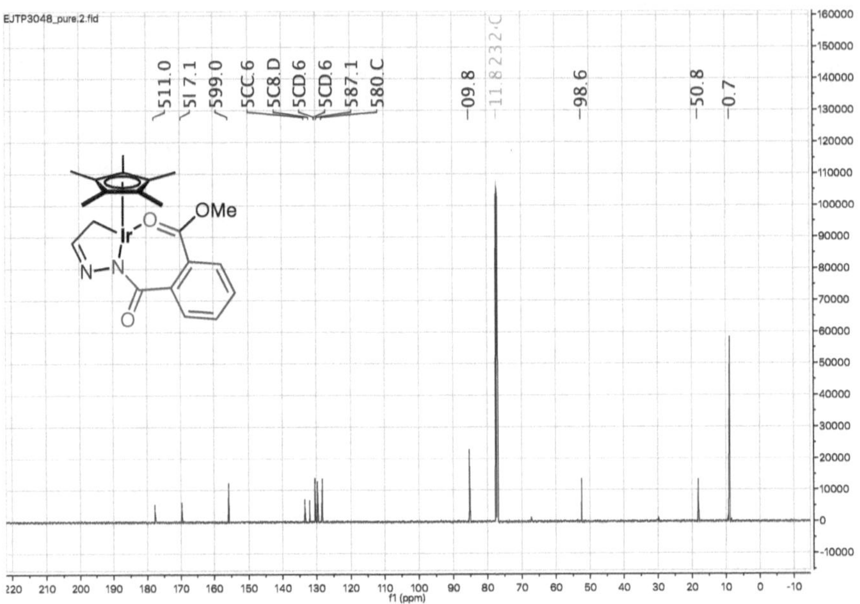

EJTP3048_pure.2.fid
OMe